The

RIEMANN HYPOTHESIS
And
PRIME NUMBER THEOREM

Comprehensive Reference, Guide and Solution Manual

Daljit S. Jandu

Infinite Bandwidth Publishing

Copyright © 2005-2006 by Daljit S. Jandu, All rights reserved
All rights reserved under Pan American and International Copyright Conventions

Publisher's Cataloging-in-Publication
(Provided by Quality Books, Inc.)

Jandu, Daljit S.
The Riemann hypothesis and prime number theorem: comprehensive reference, guide and solution manual / Daljit S. Jandu.
p. cm.
Includes index.
LCCN 2005909256
ISBN 0-9771399-0-5 (soft cover ed.)
ISBN 0-9771399-1-3 (mass market ed.)
ISBN 1-933773-00-6
ISBN 1-933773-01-4
ISBN 1-933773-02-2
ISBN 1-933773-03-0

1. Number theory. 2. Numbers, Prime. 3. Riemann hypothesis. I. Title.

QA241.J36 2006   512.7
                      QBI05-600222

This book is available at bulk sales discounted price for Universities, Government and Corporations. To receive a quote for bulk sales please visit:
http://infinitebandwidthpublishing.com
orders@infinitebandwidthpublishing.com
Call toll free (866)521-7835

This Book is dedicated to my Mother Amar Kaur and Father Teja Singh for their teaching and dedication

Contents

Preface- Note to the Reader
About the Author
Acknowledgments
Warning- Disclaimer

THE INFINITE SERIES 19

THE EULER'S FAMOUS 33

    THE GENUS FUNCTION 33

CLOSING VALUE OF APERY CONSTANT 41

THE NUMERIC WAVE EQUATION 61

THE PRIME NUMBER THEOREM 67

    THE PRIME COUNTING FUNCTION 67
    The Error or the Correction Factor 67
    APPENDIX 1 73

THE QUADRATIC RESIDUE AND EULER'S CRITERION ... 73

    APPENDIX 2 77

THE GAUSS THEOREM 77

    APPENDIX 3 85

THE PROPERTIES OF ZETA FUNCTION $\zeta(I)$ 85

    APPENDIX 4 89

FERMAT'S LITTLE THEOREM AND CHINESE REMAINDER THEOREM 89

    APPENDIX 5 93

SPECIAL FUNCTIONS 93

    APPENDIX 6 103

VALUES OF THE FIRST FEW ZETA FUNCTIONS 103
   APPENDIX 7 107
      THE COMPLEX NUMBER 107
   APPENDIX 8 113
RIEMANN ZETA FUNCTION ANALYSIS I 113
   APPENDIX 9 123
THE RIEMANN ZETA FUNCTION 123
   APPENDIX 10 145
APPENDIX 11 151
RAMANUJAN ODD ZETA FUNCTION 151
APPENDIX 12 157
AN ELLIPTIC FORM OF INFINITE SERIES 157
   APPENDIX 13 163
SOME FORMULAS CONNECTING Γ 163
   APPENDIX 14 167
ANALYSIS OF RIEMANN ZETA WITH 167
   APPENDIX 15 173
GLOSSARY 173
INDEX 183

Preface

I have seen farther than others, it is essentially because I stood on the shoulders of giants.
— Sir Isaac Newton

Periodic function frequently occurs in Information Technology and Engineering. The combination of periodic function with simple trigonometric function such as sine or cosine is the matter of great practical importance and is called Fourier series. Euler found the sum for all such possible series using infinite product on the basis of his solid judgment of factorial or remainder theorem.

The Gauss fundamental result was an easy way to sum n integers, which later became the foundation mathematical induction and computer science viz.

$$\sum_{n=1}^{k} n = \frac{n(n+1)}{2}$$

Along with the Euler's famous result viz.:

$$1 + \frac{1}{2^2} + \frac{1}{3^2} + \frac{1}{4^2} + \ldots = \frac{\pi^2}{6}$$

Euler's method of solving the closing value of the famous infinite series result can be found documented in his collected work *Opera Omnia*. The derivation is referred in Chapter 3 of this book.

The closing sum for Apery constant, a profound mystery, perhaps in the past only attempted by Indian mathematician and number theorist S. Ramanujan and yet very few thought of the mammoth real world engineering applications in information technology. The mystery that remained unraveled for centuries is revealed in this book.

The Apery constant herein means the closing sum of the infinite series:

$$1+\frac{1}{2^3}+\frac{1}{3^3}+\frac{1}{4^3}+\ldots\ldots =$$

and subsequently the sum of the series of the form for odd powers, i.e.:

$$1+\frac{1}{2^5}+\frac{1}{3^5}+\frac{1}{4^5}+\ldots\ldots =$$

Or, in general

$$\sum_{n=1}^{\infty}\frac{1}{n^{2k+1}} = \text{ for } k \in N$$

Not much work is found to have done towards Apery constant except that Apery developed a proof towards it irrationality. From IT applications point of view, concrete number theory corresponds to mathematical induction and abstract number theory corresponds to electrical induction; the mystery remained how is electrical induction is the subset of mathematical induction and vice versa and hence comes the vast application world of IT, software development, data communication, bio-informatics, computer/ network security, internet and intranet, searching and sorting and other numeric algorithms and much more.

To be more introductory to general audience, I initiate with the introduction to infinite series, the limits up to which the fundamental mathematics tools can help us to solve the closing values of converging infinite series. However, for practical limitations we shall be only interested in series going in the sequential form i.e. 1,2,3,4,........(for Bernoulli constant in relation to mathematical induction) or odd numbered series with alternate positive and negative signs (for Euler's number in relation to mathematical induction) and new definition for Euler's extension number in Appendix 15.

In this way, we shall be covering every infinite series analysis from

$Ln2 = 1 - 1/2 + 1/3 - 1/4 + \ldots$

to

$\zeta(2k+1) = \sum_{n=1}^{\infty} \frac{1}{n^{2k+1}} \text{ where } k = 1,2,3,\ldots$

Why the mathematical tools beyond the first principles are not sufficient to solve these series, what are the limitations and why not the complex algebra truly works beyond certain limits are explained in details inside this book. The author presents the proof for closing Apery constant and hence the Prime Counting Function.

In an annual conference of American Mathematical Society; the President of the society mentioned the importance of first principles, or the fundamental concepts. In his speech the President of AMS stated: "The research in mathematical science lacks fundamental essence; no wondered the students are baffled by the very concept of mathematics research. In essence, the University research in mathematics is limited in the sense, as if we are trying to perform the magical feat on the fifth floor of the buildings without considering the foundation through first four floors."

And therefore, in the following chapters we shall apply the first principles to solve the mystery. I will cheerfully pay $1.00 to the first finder of each technical, typographical, or historical error. The webpage cited on page (ii) shall contain the current listing of all corrections reported to me.

Daljit S. Jandu   December 27, 2005
North Hollywood, CA

About the Author

Daljit Jandu is inventor in varied fields of Information Technology. He holds advanced degree in Engineering from The University of Texas at Austin.

This book started just as a hobby. It didn't start as the research work in Riemann Hypothesis but with the central idea which drives the Riemann Hypothesis, viz. Apery Constant.

Being an engineer Daljit Jandu has keen interest in central idea for the scientific basis and/or direction of new technologies and his guidance and conversations are real world and based on factual mathematics.

Daljit Jandu was prompted to write this book because so many people want to have right direction towards solution of Riemann Hypothesis and the solution shall address the issues of far significance.

Acknowledgments

The following people contributed directly or indirectly to this book:

Landon Clay, Dean Vijay Dhir, Dr. Arwin Dougal, Bruce Bracken, Jim Carlson, Dan Poynter, Jan Nathan, Arnold Schmidt, Brian Jud, Wallace Prescott, Vinod Khosla, Bill Gates, Donald Knuth, Bjarne Stroustrup, Dr. Vijaya Ramachandran, John Kerrigan, Dr. Davor Juricic, Azim Premji, Dr. Panton, Dr. Gary Pope, Dr. Kant Kothawala, Dr. Aleksandar Ivic, Harold M. Edwards, John Hubbard, Julian Havil, John Derbyshire, George Andrews, Jeffrey Vaaler, William Dunham, Andrew Weil, James Ralston, Sidney McPhee

Warning- Disclaimer

The objective of the author is to educate and entertain. The author and/or Infinite Bandwidth Publishing do not assume any liability based on information contained in this book. In field, the project managers, engineers, and/or other individuals assigned to/working on the individual projects are responsible for appropriate project engineering and implementations

Notation and Symbols

Though notation for particular lemmas and theorems are presented in the respective context, the following symbols are the generally accepted notations in this text:

p, q   Prime Numbers

s   Complex variable with real and imaginary parts

Even Zetas   Refer to Appendix 15

Odd Zetas   Refer to Appendix 15

Positive Infinite Product   Refer to Appendix 15

Negative Infinite Product   Refer to Appendix 15

Boundary Conditions	Refer to Appendix 15
Euler's extension number	Refer to Appendix 15
Gauss Theorem	Refer to Appendix 2
Res $F(x)_{x=a}$	The residue of function $F(x)$ at point $x = a$; Refer to Appendix 15
x mod y	Modulus function; i.e. the remainder when x divided by y
$u(x)$ mod $v(x)$	Remainder of polynomial function u divided by polynomial function v
$\zeta(s)$	Riemann Zeta Function $$\zeta(s)=\sum_{n=1}^{\infty}\frac{1}{n^s} \ for \ \mathrm{Re}\, s > 1$$

γ	Euler's constant $$\gamma = \sum_{n=1}^{\infty} \frac{1}{n} - \ln x = 0.57$$
$\chi(s)$	The functional equation of Riemann zeta function defined in the way such that $$\zeta(s) = \chi(s)\zeta(1-s)$$ where $$\chi(s) = \frac{(2\pi)^s}{[2\Gamma(s)\cos(\pi s/2)]}$$
ρ	$\rho = \beta + i\gamma$, complex zero of $\zeta(s)$
N(T)	The number of roots $\rho = \frac{1}{2} + i\gamma$ of $\zeta(s)$ for which $0 < \gamma < T$
$\prod_{r(x)} f(x)$	The product of all f(x) such that the variable x is an integer and the relation r(x) is true

$\sum_{r(x)} f(x)$ 	The sum of all f(x) such that variable x is an integer and the relation r(x) is true

π(x) 	Number of Primes less than x

li x 	$\int_0^x \frac{dt}{\ln t} = \underset{Lim}{\Delta \to 0} \left(\int_0^{1-\Delta} \frac{dt}{\ln t} + \int_{1+\Delta}^x \frac{dt}{\ln t} \right)$

j⊥k 	Indicates that 'j' and 'k' are mutually prime g.c.d. (j,k) = 1

Φ(n) 	Euler totient function defined as the product over all prime divisor of n such that

$$\Phi(n) = n \prod_{p|n} \left(1 - \frac{1}{p}\right)$$

Refer to Appendix 15

g.d.c. 	Greatest common divisor

$\mathbb{N}$ — Set of Natural Numbers
1, 2, 3,......

$\mathbb{I}$ — Set of Integers
....-2, -1, 0, 1, 2...

$\mathbb{R}$ — Set of Real Numbers
0.32, 6/5 etc.

1

The Infinite Series

You can fool all people some of the time, and some people of all the time, but you cannot fool all the people of all the time.
                                             -Abraham Lincoln

Study as if you were to live for ever. Live as if you were to die tomorrow.
                                             - Mahatma Gandhi

Imagination is more important than knowledge
                                             - Albert Einstein

The infinite series may be of many forms but our criteria of infinite series shall be limited by the infinite series which is derived as the result sequential process or alternate sequential process and all the possible forms are included in that sequential form.

A simple infinite series is:

$1+2+3+4+\ldots\ldots\ldots\infty$ the sum of which is infinite of course but for limited number of terms the sum can be represented by simple formula, viz.

$$\frac{n(n+1)}{2}$$

If we want to sum the say first 1000 integers, we obtain $(1000 \times 1001)/2 = 500 \times 1001 = 5500$

In the similar way, we can find the sum of limited number of terms of the series for any power

$1 + 2^k + 3^k + 4^k + \ldots\ldots\ldots$

Where $k = 1, 2, 3\ldots$ Or, $k \in \mathbb{N}$

In such sum we can conveniently introduce Bernoulli's number B_n. Later in Appendix 12 we find different values of this Bernoulli's number.

However, things are little different in the reciprocal power,

$$1 + \frac{1}{2^k} + \frac{1}{3^k} + \frac{1}{4^k} + \ldots \ldots \ldots \quad (1)$$

Where k=1, 2, 3...Or, k ∈ ℕ

For simple k=1, the harmonic series diverges

$$1 + \frac{1}{2} + \frac{1}{3} + \frac{1}{4} + \ldots \ldots \ldots \infty = \infty$$

It can be proved by grouping two terms followed by four terms followed by eight terms and so on.

For example $1/3 + 1/4 > ½$

Similarly, $1/5 + 1/6 + 1/7 + 1/8 > ½$

$1/9 + 1/10 + 1/11 + 1/12 + 1/13 + 1/14 + 1/15 + 1/16 > ½$

In this way, proceeding to infinity, we conclude that the sum of this harmonic series for infinity number of terms is infinity or the above harmonic series is diverging.

What happens to series (1) for k>1, where k is a real number or, k ∈ℝ. Of course for any integer greater than 1, i.e. 2, 3, 4...the series type (1) converges and for even powers to a closing value in terms of π.

For example,

$$\zeta(2) = 1 + \frac{1}{2^2} + \frac{1}{3^2} + \frac{1}{4^2} + \ldots = \frac{\pi^2}{6}$$

Or, in general

$$\zeta(2n) = \frac{2^{2n-1} \pi^{2n} B_n}{(2n)!} \quad n = 1, 2, 3, \ldots$$

The mystery remained for the closing value of $\zeta(2n+1)$ where n = 1, 2, 3 ...Or, the mystery abounding odd zetas. S. Ramanujan did some work towards the closing values for the odd powers of such series as listed in Appendix 11. Is there an error in Ramanujan's work? As Hardy, an English mathematician pointed out that Ramanujan's work could only be from the mathematician of highest merit. But Ramanujan's derivation assumes the value of e=2.71...to be a universal constant like the value of π.

There is a big difference between universal constant (i.e. the value of π = 3.14159.....) and the conventional constant (i.e. the value of e=2.71......) as explained in later chapters.

The Universal Constant π follows the cyclicity and hence the Euler and Wilson theorem as well as Chinese Remainder theorem and therefore is quite different than e=2.71...

Let's define e= 2.71...by Taylor series

$$e^x = 1 + \frac{x}{1!} + \frac{x^2}{2!} + \frac{x^3}{3!} + \ldots \ldots \quad (2)$$

Substituting x = ix in (2) one gets

$$e^{ix} = \left[1 - \frac{x^2}{2!} + \frac{x^4}{4!} - \ldots \ldots \right] + i\left[x - \frac{x^3}{3!} + \frac{x^5}{5!} - \ldots \ldots \right]$$

Or,

$$e^{ix} = \cos x + i \sin x \quad (3)$$

The equation (3) was discovered by Euler, and he wrote that this equation has to be true but he does not understand what does this equation means.

The famous Euler's results viz. ζ(2), ζ(4), ζ(6)...were

basically derived from the premise of roots of equation $\sin x = 0$ in the factorial form

$$\sin x = \prod_{n=1}^{\infty}\left[1 - \frac{x^2}{(n\pi)^2}\right]$$

The Euler method of computing the even powered zeta function can be extended to the equation $\cos x = 0$ or in general to the equation

$$[a.\cos x + b.\sin x] = 0 \tag{4}$$

In fact the equation (4) can be reduced to

$$\frac{a}{\sqrt{a^2+b^2}}\cos x + \frac{b}{\sqrt{a^2+b^2}}\sin x = 0 \tag{5}$$

Now, if

$$\sin \alpha = \frac{a}{\sqrt{a^2+b^2}} \qquad \cos \alpha = \frac{b}{\sqrt{a^2+b^2}}$$

Then the equation (5) becomes

$$\sin(x + \alpha) = 0 \tag{6}$$

This too generate even powered infinite series.

Are the equation (3) and (4) related? Not really. First, because equation (3) does not has a phase difference as equation (6). Second, because the equation (3) being a unit circle cannot be equated to zero. Third and most importantly Euler's result i.e. $\zeta(2n)$ where n ∈ ℕ were derived from the premise that positive and negative go together sequentially, the major characteristic related to π which is absent in e.

Now, let

$$e^x = n \xrightarrow{Lim} \infty \left(1 + \frac{x}{n}\right)^n \qquad (6a)$$

Hence

$$e^{-x} = n \xrightarrow{Lim} \infty \left(1 - \frac{x}{n}\right)^n \qquad (6b)$$

Multiplying and normalizing equations (6a) and (6b); one obtains the sine product formula. As per global algorithmic design we can efficiently substitute 1 for 0 and 0 for 1.

$$\prod_{n=1}^{\infty}\left(1-\frac{x^2}{n^2}\right) = 0 \: or \: 1 = Sin(\pi x) \tag{7}$$

Equation (6a), (6b) and (7) explains why most of the results related to Riemann Hypothesis in complex theory are mere approximations while in real analysis are true.

Though the functional equation developed by Riemann (discussed in later chapters) is considered true for every point except in the trivial points -2, -4, -6…justifies the equation unsuitable for approximations of the hypothesis for number of primes less than given number. This is so because the triviality can subtly extend to odd zetas viz.

$$(-1)\left(\frac{1}{2^3}+\frac{1}{4^3}+\frac{1}{6^3}+\ldots\ldots\right) = \left(-\frac{1}{2^3}\right)\left(1+\frac{1}{2^3}+\frac{1}{3^3}+\frac{1}{4^3}+\ldots\ldots\right)$$

Similarly,

$$(-1)\left(\frac{1}{2^5}+\frac{1}{4^5}+\frac{1}{6^5}+\ldots\ldots\right) = \left(-\frac{1}{2^5}\right)\left(1+\frac{1}{2^5}+\frac{1}{3^5}+\frac{1}{4^5}+\ldots\ldots\right)$$

$$(-1)\left(\frac{1}{2^7}+\frac{1}{4^7}+\frac{1}{6^7}+\ldots\ldots\ldots\right)=$$
$$\left(-\frac{1}{2^7}\right)\left(1+\frac{1}{2^7}+\frac{1}{3^7}+\frac{1}{4^7}+\ldots\ldots\ldots\right)$$

And so on.

The triviality of Riemann's functional equation leaves the same question; i.e. the closing value of odd zeta or, $\zeta(3)$, $\zeta(5)$, $\zeta(7)$,........

The closing value of even zetas $\zeta(2)$, $\zeta(4)$, $\zeta(6)$...are well established in relation to the even powers of π. It is closing values of even zeta function i.e. $\zeta(2)$, $\zeta(4)$, $\zeta(6)$, $\zeta(8)$,...........which gives power of it's equivalence to Wilson Theorem and Fermat's Little Theorem. For Wilson Theorem or Fermat's Little Theorem please refer to Appendix 4.

Another point of importance is as why the convergence of Riemann Hypothesis is around ½; i.e. as why the real part of non-trivial zero is ½.

This can be explained in terms of two unique infinite series which are of central importance of the boundary values in relation to Riemann Hypothesis. Let us reconsider the

28 -Riemann Hypothesis and PNT

infinite product of $\sin(\pi x)$ i.e.,

$$\sin(\pi x) = \prod_{n=1}^{\infty}\left(1 - \frac{x^2}{n^2}\right)$$

Putting x=1, the first term of the product is 0; setting it to 1 for global consideration. Therefore, the product becomes:

$$\left(1-\frac{1}{2^2}\right)\left(1-\frac{1}{3^2}\right)\left(1-\frac{1}{4^2}\right)\ldots\ldots\ldots = \frac{1}{2}$$

Another infinite product representing the boundary conditions, i.e.

$$\left(1-\frac{1}{2}\right)\left(1+\frac{1}{3}\right)\left(1-\frac{1}{4}\right)\left(1+\frac{1}{5}\right)\ldots\ldots = \frac{1}{2} \qquad (8a.)$$

And it's complementary series:

$$\left(1+\frac{1}{2}\right)\left(1-\frac{1}{3}\right)\left(1+\frac{1}{4}\right)\left(1-\frac{1}{5}\right)\ldots\ldots = 1 \qquad (8b.)$$

In fact the product form of $\sin(\pi x)$ [or $\sin(x)$] can be written as the product of two complementary function; if one such function is F(x) the other function shall obviously be F(-x).

F(x) and F(-x) may not necessarily be the only two complementary function represented above strictly in terms of sign. In fact the sign can take any form to the maximum boundary value.

For example if one function is at the extreme boundary i.e.

$$F(x) = \left(1+\frac{1}{1}\right)\left(1+\frac{1}{2}\right)\left(1+\frac{1}{3}\right)\left(1+\frac{1}{4}\right)\dots\dots$$

then the other function shallbe

$$F(-x) = \left(1-\frac{1}{2}\right)\left(1-\frac{1}{3}\right)\left(1-\frac{1}{4}\right)\dots\dots$$

(9)

The first product term of F(-x) which is 0 is set to 1 to make function global. In fact this F(x) and F(-x) form represent one of the extreme boundary conditions. What would be other extreme boundary conditions?

Obviously, in the form:

$$F(x) = \left(1+\frac{1}{1}\right)\left(1-\frac{1}{2}\right)\left(1+\frac{1}{3}\right)\left(1-\frac{1}{4}\right)\dots\dots = 1$$

and

$$F(-x) = \left(1+\frac{1}{2}\right)\left(1-\frac{1}{3}\right)\left(1+\frac{1}{4}\right)\dots\dots = 1$$

(10)

The first product term of F(-x) is 0 and is set to 1 for global consideration. The boundary conditions by the set of equations (9) and (10) encompasses every possible combinations from sign criteria and with this boundary condition we shall prove the unsolved mystery of odd zetas i.e. the mystery surrounding the closing value of $\zeta(3)$, $\zeta(5)$, $\zeta(7)$, $\zeta(9)$,............ The boundary conditions encompassed in the set of complementary equations (9) and (10) constitute the boundary conditions of any quantum system.

In general,

$$\sin x = F(x).F(-x) = 0 \; or \; 1 \tag{11}$$

The equation (11) presents a vague scenario. Are there non-sinusoidal periodic function? If yes, how can be they applied to vast variety of engineering and allied field e.g. engineering analysis, electrical engineering and computer science and data communication? The entire Fourier Transformation and Series shall be the subset of the solution in this scenario. The gamma function which serves as the basis for most computations regarding Riemann Functional equation and the non-trivial roots in the form of ½ + bi is the mere representation of the equation type (11) and it does not take into consideration all the scenario of random positives and negatives which would be impossible to consider unless boundary condition basis is called upon to make such computation feasible.

How is the solving of this scenario leads to the solution to the closing values of odd zetas viz. $\zeta(3)$, $\zeta(5)$, $\zeta(7)$,.....................?

As is noted that the real part of non trivial zeros of zeta function is ½ is self-explanatory from the equation 8a and 8b. In case of complex analysis for the non-trivial roots of simple complex function

$$e^{ix} = \cos x + i \sin x \qquad (12)$$

The equation 12 represent a unit circle in complex plane, but not in Cartesian plane (in number theory).

In number theory, the unit circle shall be represented by substituting x by -x in the equation (12) and then multiplying the equation (12) by the result i.e.

$$e^{ix} \cdot e^{-ix} = [\cos x + i \sin x][\cos x - i \sin x] = \cos^2 x + \sin^2 x = 1$$

This represents a unit circle. At the global level, in the number theory, the positive and negative infinite products go together and the unilateral positive or negative infinite products lead to erroneous results. In the following chapters we shall discuss the Euler's Famous result and prove the most important result in the history of mathematics, the closing value of Apery constant and odd zetas viz. the closing values of $\zeta(3)$, $\zeta(5)$, $\zeta(7)$,......

2

The Euler's Famous Result

The Genus Function

The definition of God and Man

The man is infinite circle whose circumference is nowhere, but the center is located in one spot and God is an infinite circle whose circumference is nowhere, but whose center is everywhere.

-Swami Vivekanand

The function

$$\sin(x) = x \prod_{n=1}^{\infty} \left[1 - \frac{x^2}{(n\pi)^2} \right]$$

Or, in a more simplified way

For $\sin(x) = 0$ and $x \neq 0$; yields

34 - Riemann Hypothesis and PNT

$$\prod_{n=1}^{\infty}\left[1-\frac{x^2}{(n\pi)^2}\right] = 1 - \frac{x^2}{3!} + \frac{x^4}{5!} - \frac{x^6}{7!} + \ldots \infty \qquad (1)$$

It may be noted that (1) is obtained by using Factorial or Remainder theorem i.e. $\sin(x) = 0$ gives the roots $x = \pm\pi, \pm 2\pi, \pm 3\pi, \pm 4\pi\ldots$

$$\left[1-\frac{x^2}{\pi^2}\right]\left[1-\frac{x^2}{4\pi^2}\right]\left[1-\frac{x^2}{9\pi^2}\right]\ldots = 0$$

$$= 1 - \left\{\frac{1}{\pi^2} + \frac{1}{4\pi^2} + \frac{1}{9\pi^2} + \ldots\right\}x^2 + \{\ldots\}x^4 - \{\ldots\}x^6 + \ldots \qquad (2)$$

Comparing the coefficient of x^2 in the equation (2) gives the Euler's famous result of 1734:

$$\sum_{n=1}^{\infty}\frac{1}{n^2} = \frac{\pi^2}{6}$$

In other words,

$$1 + \frac{1}{2^2} + \frac{1}{3^2} + \frac{1}{4^2} + \frac{1}{5^2} + \frac{1}{6^2} + \ldots \infty = \frac{\pi^2}{6} \qquad (3)$$

Theorem

Lejeune Dirichlet:

Abhandlungen Koniglich Preuβ Akad (1849)

The probability of picking two random numbers such that they have no common factor or their g.c.d. =1 is $\frac{6}{\pi^2}$

If x and y are integers chosen at random, the probability that g.d.c. (x, y) =1 is

$$\frac{6}{\pi^2} = 0.60793$$

Let us assume that g.d.c. (x, y) = d; which is only possible if x is the multiple of d and y is the multiple of d.

Or, x/d ⊥y/d and let us assume the probability p for x⊥y, or g.d.c. (x, y) =1

Therefore, for all integers we shall obtain:

$$\sum_{d=1}^{\infty} \frac{p}{d^2} = p\left[1 + \frac{1}{2^2} + \frac{1}{3^2} \ldots \ldots \infty \text{ terms}\right] = 1$$

Or,

$$p = \frac{6}{\pi^2}$$

QED

From the series (3) we can easily derive

$$1 + \frac{1}{3^2} + \frac{1}{5^2} + \frac{1}{7^2} + \ldots \ldots \infty = \frac{\pi^2}{6} - \frac{1}{2^2}\left\{1 + \frac{1}{2^2} + \frac{1}{3^2} + \frac{1}{4^2} + \frac{1}{5^2} + \ldots\right\} = \frac{\pi^2}{12}$$

In fact for every n we have this relationship:

$$\sum_{k=0}^{\infty} \frac{1}{(2k+1)^n} = \left[1 - \frac{1}{2^n}\right] \sum_{k=1}^{\infty} \frac{1}{k^n}$$

By simple algebraic extensions Euler was able to prove

$$1 + \frac{1}{2^4} + \frac{1}{3^4} + \frac{1}{4^4} + \frac{1}{5^4} \ldots \ldots \infty = \frac{\pi^4}{90}$$

$$1 + \frac{1}{2^6} + \frac{1}{3^6} + \frac{1}{4^6} + \frac{1}{5^6} \ldots \ldots \infty = \frac{\pi^6}{945}$$

………………. All the way to

$$1+\frac{1}{2^{26}}+\frac{1}{3^{26}}+\frac{1}{4^{26}}+\frac{1}{5^{26}}+\ldots\ldots\infty = \frac{1315862}{11094481976030578125}\pi^{26}$$

In a general case this finding of closing value for even zetas can go up to infinite power and is given by equation:

$$\frac{1}{1^{2p}}+\frac{1}{2^{2p}}+\frac{1}{3^{2p}}+\frac{1}{4^{2p}}+\frac{1}{5^{2p}}+\ldots\ldots\infty = \frac{2^{2p-1}\pi^{2p}Bp}{(2p)!}$$

Where Bp is the Bernoulli number

Bernoulli numbers ......are defined by the series

$$\frac{x}{e^x-1} = 1-\frac{x}{2}+\frac{B_1 x^2}{2!}-\frac{B_2 x^4}{4!}+\frac{B_3 x^6}{6!}-\ldots\ldots$$

$$1-\frac{x}{2}\cot\frac{x}{2} = \frac{B_1 x^2}{2!}+\frac{B_2 x^4}{4!}+\frac{B_3 x^6}{6!}+\ldots\ldots$$

$$|x| \leq \pi$$

Series Involving Bernoulli Numbers:

$$B_n = \frac{(2n)!}{2^{2n-1}\pi^{2n}}\left[1 + \frac{1}{2^{2n}} + \frac{1}{3^{2n}} + \ldots\ldots\right]$$

$$B_n = \frac{2(2n)!}{(2^{2n}-1)\pi^{2n}}\left[1 + \frac{1}{3^{2n}} + \frac{1}{5^{2n}} + \ldots\ldots\right]$$

$$B_n = \frac{2(2n)!}{(2^{2n-1}-1)\pi^{2n}}\left[1 - \frac{1}{2^{2n}} + \frac{1}{3^{2n}} - \ldots\ldots\right]$$

The Euler numbers are defined by the series

$$\sec h(x) = 1 - \frac{E_1 x^2}{2!} + \frac{E_2 x^4}{4!} - \frac{E_3 x^6}{6!} + \ldots\ldots$$

$$\sec(x) = 1 + \frac{E_1 x^2}{2!} + \frac{E_2 x^4}{4!} + \frac{E_3 x^6}{6!} \ldots\ldots$$

$$|x| \leq \frac{\pi}{2}$$

Series Involving Euler Numbers:

$$E_n = \frac{2^{2n+2}(2n)!}{\pi^{2n+1}}\left\{1 - \frac{1}{3^{2n+1}} + \frac{1}{5^{2n+1}} - \frac{1}{7^{2n+1}} + \ldots\ldots\right\}$$

Relationships of Bernoulli and Euler Numbers

$$\binom{2N+1}{2}2^2 B_1 - \binom{2n+1}{4}2^4 B_2 + \binom{2n+1}{6}2^6 B_3 - \ldots (-1)^{n-1}(2n+1)2^{2n} B_n = 2n$$

$$E_n = \binom{2n}{2}E_{n-1} - \binom{2n}{4}E_{n-2} + \binom{2n}{6}E_{n-3} - \ldots (-1)^n$$

$$B_n = \frac{2n}{2n(2^{2n}-1)}\left[\binom{2n-1}{1}E_{n-1} - \binom{2n-1}{3}E_{n-2} + \binom{2n-1}{5}E_{n-3} - \ldots (-1^{n-1})\right]$$

We note that Euler's famous results and all the even powered series are in the field of Bernoulli's number. Euler result is called genus function because it serves as the basis for the closing value of all the even powered series and subsequently for Apery constant and all the odd powered series (Chapter 3); while the Euler's numbers have implication only on particular type alternate sign series viz.:

$$1 - \frac{1}{3^3} + \frac{1}{5^3} - \ldots \infty \text{ terms}$$

$$1 - \frac{1}{3^5} + \frac{1}{5^5} - \ldots \infty \text{ terms}$$

$$1 - \frac{1}{3^7} + \frac{1}{5^7} - \ldots \infty \text{ term}$$

and so on.......

And therefore the number associated with other type of series, e.g. series (4) shall be called Euler's extension numbers

$$1 - \frac{1}{7^3} + \frac{1}{9^3} - \frac{1}{15^3} + \ldots\ldots\ldots \infty \, terms \qquad (4)$$

It is worthwhile to note that there are infinite Euler's extension numbers.

3

Closing Value of Apery Constant

The grand aim of all science is to cove the greatest number of empirical facts by logical deduction from smallest number of hypotheses and axioms .

- Albert Einstein

Those who see action, where there is inaction and inaction, where there is action are wise men

- Bhagavad-Gita (6th. Century B.C.)

The Euler's famous results for even zetas are well known for their design, elegance and beauty, i.e. $\zeta(2), \zeta(4), \zeta(6), \zeta(8)$...

42 -Riemann Hypothesis and PNT

In general $\zeta(2n)$ for n=1, 2, 3... is represented as

$$\zeta(2n) = \frac{2^{2n-1} \pi^{2p} B_{2n}}{(2n)!}$$

Where B_{2n} is the Bernoulli's number

It has been known conjectures for two centuries that $\zeta(2n+1)$ where n= 1,2,3,....... can be represented in the form

$$\zeta(2n+1) = \sum_{N=1}^{\infty} \frac{1}{N^{2n+1}} = \frac{p}{q} \pi^{2n+1}$$
for every $n = 1,2,3,4,........$

Where p and q are positive integers

It just meant

$$\zeta(5) = \sum_{N=1}^{\infty} \frac{1}{N^5} = 1 + \frac{1}{2^5} + \frac{1}{3^5} + \ldots \ldots \ldots \infty \, terms = \frac{p_1}{q_1} \pi^5$$

Where $p_1, q_1 \in \mathbb{N}$

The conjecture was assumed for every odd zeta, viz. $\zeta(7), \zeta(8), \zeta(9),\ldots\ldots$ And so on.

Why we have the reason to believe that above equations for the $\zeta(2n+1)$ are not correct?

Let us consider a simple polynomial of infinite degree, viz.

$$F(x) = 1 + x + x^2 + x^3 + x^4 + \ldots\ldots x^\infty \qquad (1)$$

Then we have,

$$F(-x) = 1 - x + x^2 - x^3 + x^4 + \ldots\ldots x^\infty \qquad (2)$$

Multiplying (1) and (2) we obtain:

$$F(x).F(-x) = 1 + x^2 + x^4 + x^6 + \ldots\ldots + x^\infty \qquad (3)$$

From equation (3) it's obvious that the only comparison of powers we can do are even powers which justifies and confirms Euler's findings and of the product formulation for sine and other trigonometric functions e.g.,

$$\sin(\pi x) = \prod_{n=1}^{\infty}\left[1 - \frac{x^2}{n^2}\right] \tag{4}$$

Similarly,

$$\cos\left(\frac{\pi x}{2}\right) = \prod_{n=1}^{\infty}\left[1 - \frac{x^2}{(2n+1)^2}\right] \tag{5}$$

It shall be fascinating to see that the equation (5) yields to the sum of series:

$$1 + \frac{1}{3^2} + \frac{1}{5^2} + \frac{1}{7^2} + \dots \infty \text{ terms} = \frac{\pi^2}{12}$$

And all series of the form

$$\sum_{n=1}^{\infty}\frac{1}{(2k+1)^n} = \left[1 - \frac{1}{2^n}\right]\sum_{k=1}^{\infty}\frac{1}{k^n} \text{ where } k = 1,2,3,\dots \tag{6}$$

Which is the same as by Euler's method of finding the closing value ζ(2), ζ(4), ζ(6)... In fact

$$1 + \frac{1}{2^2} + \frac{1}{3^2} + \frac{1}{4^2} + \dots \infty \text{ terms}$$

$$= 1 + \frac{1}{3^2} + \frac{1}{5^2} + \frac{1}{7^2} + \dots \infty \text{ terms} + \frac{1}{2^2}\left[1 + \frac{1}{2^2} + \frac{1}{3^2} + \frac{1}{4^2} + \dots \infty \text{ terms}\right]$$

$$= \frac{\pi^2}{8} + \frac{\pi^2}{24} = \frac{\pi^2}{6}$$

Chapter 3 Closing Value of Apery Const.- 45

Similarly the relationship for conversion of the closing values for all the even power exists.

Interestingly, since

$[\cos(x) - \sin(x)][\cos(x) + \sin(x)] = \cos(2x)$ we have,
$\cos(\pi x/2) = [\cos(\pi x/4) - \sin(\pi x/4)][\cos(\pi x/4) + \sin(\pi x/4)]$

The roots of $\cos(x) - \sin(x) = 0$ can be found and compared to polynomial

$$\cos x - \sin x = 1 - x - \frac{x^2}{2!} + \frac{x^3}{3!} + \frac{x^4}{4!} - \frac{x^5}{5!} - \frac{x^6}{6!}\ldots\ldots\ldots$$

$$\cos x - \sin x = \sqrt{2}\left[\frac{1}{\sqrt{2}}\cos x - \frac{1}{\sqrt{2}}\sin x\right] = 0$$

$$\cos\left(\frac{\pi}{4} + x\right) = 0$$

hence the roots are
$\pm \pi/4 \,;\pm\, ^3\pi/4\,;\pm\, ^5\pi/4\,,\ldots\ldots\ldots$

Similarly,

$\cos(x) + \sin(x) = 0$ yields,

$$\cos\left[x - \tfrac{\pi}{4}\right] = 0$$

Which has the same roots as cos(x) - sin(x) = 0

Using the Factorial (or Remainder) theorem generates equation (5)
Or, we can simply say:

cos (x) = F(x). F(-x)

It's worth noting the above complementary functions for cos (x). Again there are infinite numbers of such complementary function possible, e.g.

$\cos x - \sqrt{3}\sin x$ and $\cos x + \sqrt{3}\sin x$

This can be expressed as:

$\cos(x + \pi/3)$ and $\cos(x - \pi/3)$

Or,

$\sin(x - \pi/6)$ and $\sin(x + \pi/6)$

In general a.cos(x) ± b.sin can be expressed as

$\cos(x \pm \alpha)$ or $\sin(x \pm \beta)$

Therefore, it is easily concluded that any sine or cosine function with phase difference represents two complementary functions and generate the even powered odd numbered series. The periodicity of odd numbered series depends upon phase difference. Since there is always a possibility of rational number multiple of π as the phase difference, there are infinite number of even powered odd numbered series.

To extend the results to odd powered odd numbered series,

$\cos(x) = F(ax) \cdot F(-ax)$ \hfill (7)

where $F(x) = \left(1 - \dfrac{x}{3}\right)\left(1 + \dfrac{x}{5}\right)\left(1 - \dfrac{x}{7}\right)\ldots$

and $F(-x) = \left(1 + \dfrac{x}{3}\right)\left(1 - \dfrac{x}{5}\right)\left(1 + \dfrac{x}{7}\right)\ldots$ \hfill (8)

To find the closing value of cubic series, we multiply both sides of (7) by F(-x) or F(x) and the resultant series is obtained:

$$1-\frac{1}{3^3}+\frac{1}{5^3}-\frac{1}{7^3}+\ldots\ldots\ldots\ldots\infty\ terms = \frac{\pi^3}{35} \quad (9)$$

The closing value of series (9) can be derived by many methods. There are infinite numbers of such series with alternate positive and negative signs.

Similarly,

$$1-\frac{1}{3^5}+\frac{1}{5^5}-\frac{1}{7^5}+\ldots\ldots\ldots\ldots\infty\ terms = \frac{p_1}{q_1}\pi^5 \quad (10)$$

Where p_1, q_1 are positive integers

Or,

$$1-\frac{1}{3^7}+\frac{1}{5^7}-\frac{1}{7^7}+\ldots\ldots\ldots\ldots\infty\ terms = \frac{p_2}{q_2}\pi^7 \quad (11)$$

Where p_2, q_2 are positive integers

Chapter 3 Closing Value of Apery Const.- 49

And so on.......

Series of type (9), (10) and (11).... are controlled by or related to Euler's Number.

Euler numbers precisely control the series of the form:

$$\sum_{n=0}^{\infty} \frac{(-1)^n}{(2n+1)^{2N+1}} \quad for\ N = 1,2,3,4,\dots\dots \tag{12}$$

As there are infinite numbers of periodic odd powered infinite series with alternate negative and positive terms are possible, Euler numbers represent only the ones of the type (12).

However, the odd powered all positive terms series may not be feasibly summed to closing value unless some boundary value condition is applied. The straight knowledge based solution to the following odd powered all positive term is not known without implementing the boundary condition; for example

$$1+\frac{1}{3^3}+\frac{1}{5^3}+\frac{1}{7^3}+\frac{1}{9^3}+\ldots\ldots\ldots\infty\ terms=? \qquad (14)$$

Or, in other words the closing value of any series which is odd numbered and odd powered is not possible for straight positive signs. However, with alternate positive and negative signs infinite many periodicities are possible based on the representation of $a.\cos(x) + b.\sin(x)$ as $\cos(x \pm \alpha)$ or $\sin(x \pm \beta)$.

In fact the equation (14) is the straight representation of the Apery Constant by simple extension, i.e.:

$$\sum_{n=1}^{\infty}\frac{1}{(2n+1)^3}=1+\frac{1}{3^3}+\frac{1}{5^3}+\ldots\ldots\ldots\infty\ terms$$

$$=1+\frac{1}{2^3}+\frac{1}{3^3}+\frac{1}{4^3}+\frac{1}{5^3}+\ldots\ldots\infty\ terms-\frac{1}{2^3}\left[\sum_{n=1}^{\infty}\frac{1}{n^3}\right]$$

$$=\frac{7}{8}\sum_{n=1}^{\infty}\frac{1}{n^3}$$

Similarly,

Chapter 3 Closing Value of Apery Const.- 51

$$\sum_{n=1}^{\infty}\frac{1}{(2n+1)^5} = \frac{31}{32}\sum_{n=1}^{\infty}\frac{1}{n^3}$$

Or in general

$$\sum_{n=1}^{\infty}\frac{1}{(2n+1)^N} = \frac{(2^N-1)}{2^N}\sum_{n=1}^{\infty}\frac{1}{n^N}$$

Why the summation of equation (14) is not possible from the roots of cos (x)? It is simply because

$$\cos(x) = F(x).F(-x) \qquad (15)$$

This yields simple odd numbered and even powered series of the form of equation (6).

To extend to the cubic or any odd powered series we need to multiply both sides equation (15) by either F(x) or F(-x).
Since F(x) or F(-x) are composed of terms which are alternatively positive and negative, their product yields the series with odd numbered alternate positive and negative periodic series and hence where the Euler number (and Euler's extension numbers, Appendix 15) comes into control.

The objective to find the closing value of odd powered zeta function so far remained a mystery for which we need the sequential series devoid of alternate positive and negative terms of odd powered and odd numbered series.

As we have noticed that $\cos(x) = 0$; though closely related to the results of $\sin(x) = 0$, cannot solve the closing values for the series of type

$$\sum_{n=1}^{\infty} \frac{1}{(2n+1)^N}$$

Or,

$$\sum_{n=1}^{\infty} \frac{1}{n^N}$$

where N is an odd number

The following alternative positive and negative term series, if summed in closed form shall solve our problem for finding the closing value of odd powered series;

Chapter 3 Closing Value of Apery Const.- 53

$$1 - \frac{1}{2^3} + \frac{1}{3^3} - \frac{1}{4^3} + \ldots \ldots \infty \text{ terms}$$

Or in general the alternate positive and negative term series of form

$$1 - \frac{1}{2^{(2p+1)}} + \frac{1}{3^{(2p+1)}} - \frac{1}{4^{(2p+1)}} + \ldots \ldots \infty \text{ terms}$$

Or in general form:

$$\sum_{n=1}^{\infty} \frac{(-1)^{n+1}}{p^{(2p+1)}}$$

where $p = 1, 2, 3, \ldots$

The even powered sequential series are generated by the equation $\sin(x) = 0$ as is obvious from Euler's Famous Result.

Now let us see how does the solution to the equation $\sin x = 0$ unravels the mysteries underlying the closing value of odd zetas.

Let us consider the most general and simple equation viz.

$$\sin \pi x = F(x).F(-x) = \prod_{n=1}^{\infty} \left[1 - \frac{x^2}{n^2} \right] \qquad (16)$$

In expanded form, if

$$F(x) = \left(1 - \frac{x}{1}\right)\left(1 + \frac{x}{2}\right)\left(1 - \frac{x}{3}\right)\left(1 + \frac{x}{4}\right)\ldots\ldots\infty \text{ terms} \tag{17a}$$

Then

$$F(-x) = \left(1 + \frac{x}{1}\right)\left(1 - \frac{x}{2}\right)\left(1 + \frac{x}{3}\right)\left(1 - \frac{x}{4}\right)\ldots\ldots\infty \text{ terms} \tag{17b}$$

In reality there are virtually infinite numbers of such combinations possible. Let's consider the few scenarios.

Combination 1

If

$$F(x) = \left(1 + \frac{x}{1}\right)\left(1 + \frac{x}{2}\right)\left(1 + \frac{x}{3}\right)\left(1 + \frac{x}{4}\right)\ldots\ldots\infty \text{ terms} \tag{18a}$$

Then

$$F(-x) = \left(1 - \frac{x}{1}\right)\left(1 - \frac{x}{2}\right)\left(1 - \frac{x}{3}\right)\left(1 - \frac{x}{4}\right)\ldots\ldots\infty \text{ terms} \tag{18b}$$

Combination 2

If

Chapter 3 Closing Value of Apery Const.- 55

$$F(x) = \left(1+\frac{x}{1}\right)\left(1+\frac{x}{2}\right)\left(1-\frac{x}{3}\right)\left(1-\frac{x}{4}\right)\left(1+\frac{x}{5}\right)\left(1+\frac{x}{6}\right)\ldots\ldots\ldots\infty \text{ terms}$$

Then

$$F(-x) = \left(1-\frac{x}{1}\right)\left(1-\frac{x}{2}\right)\left(1+\frac{x}{3}\right)\left(1+\frac{x}{4}\right)\left(1-\frac{x}{5}\right)\left(1-\frac{x}{6}\right)\ldots\ldots\ldots\infty \text{ terms}$$

And so on

Therefore, literally there are infinite numbers of the combinatorial possibilities for F(x) and F(-x) function as application to equation (16).

But our interest is only in the boundary conditions. A close look to 17a and 17b reveals to be one side of the boundary and 18a and 18b reveals to be on the other side. We assign the equations 17a (and 17b) as well as 18a (and 18b) to be the boundary value of our interest. They are called boundary value because they encompass infinite number of combinations for F(x) and F(-x) in relation to the sine product formula exhibited in equation (16).

The value of the sine product equation (16) is just zero. For global consideration, setting it to 1 yields

$$F(x) \cdot F(-x) = 1 \qquad (19)$$

The solution to equation (19) is:
Either
(1) Both function are 0 i.e. $F(x) = 0$ and $F(-x) = 0$

Or, (2) either of the function is zero an the other one is 1 or -1

Or, (3) both functions are -1

Or, (4) both function are unity, i.e. $F(x) = 1$ and $F(-x) = 1$

The fourth condition is most general global solution. Let us consider the one side of the boundary condition and setting it unity i.e. setting the equation 17a to unity, we get:

$$F(x) = \left(1 - \frac{x}{1}\right)\left(1 + \frac{x}{2}\right)\left(1 - \frac{x}{3}\right)\left(1 + \frac{x}{4}\right)\ldots\ldots\infty \, terms = 1 \qquad (20)$$

Using the formulas

$$(1-x^3) = (1-x)(1+x+x^2)$$
$$(1+x^3) = (1+x)(1-x+x^2)$$

Multiplying both sides of equation (20) in appropriate way, and simplifying one obtains:

$$\left[\left(1-\frac{x}{1}\right)\left(1+\frac{x}{1}+\frac{x^2}{1^2}\right)\right]\left[\left(1+\frac{x}{2}\right)\left(1-\frac{x}{2}+\frac{x^2}{2^2}\right)\right]\left[\left(1-\frac{x}{3}\right)\left(1+\frac{x}{3}+\frac{x^2}{3^2}\right)\right]\ldots\ldots =$$

$$\left(1+\frac{x}{1}+\frac{x^2}{1^2}\right)\left(1-\frac{x}{2}+\frac{x^2}{2^2}\right)\left(1+\frac{x}{3}+\frac{x^2}{3^2}\right)\left(1-\frac{x}{4}+\frac{x^2}{4^2}\right)\ldots\ldots$$

Simplifying and comparing coefficients of x^3 one obtains

$$\left(1-\frac{x^3}{1^3}\right)\left(1+\frac{x^3}{2^3}\right)\left(1-\frac{x^3}{3^3}\right)\left(1+\frac{x^3}{4^3}\right)\ldots\ldots =$$
$$\alpha_1 x + \alpha_2 x^2 + \alpha_3 x^3 \ldots \ldots \infty \text{ terms} \qquad (21)$$

Where α_1, α_2, α_3...are the coefficients of x, x^2, x^3... respectively.

Comparing the coefficients of x^3 on both sides of equation (21), one obtains:

$$1 - \frac{1}{2^3} + \frac{1}{3^3} - \frac{1}{4^3} + \ldots\ldots = |\alpha_3| + \in_1 \qquad (22)$$

Where α_3 is the coefficient of x^3 on the right side of equation (21) and ϵ_1 is the error caused by unaccounted comparison of powers of x other than the cubic power on the right hand and left hand side of equation (21)

Now let us consider the other side of boundary condition, viz., the equation 18a

$$F(x) = \left(1+\frac{x}{1}\right)\left(1+\frac{x}{2}\right)\left(1+\frac{x}{3}\right)\left(1+\frac{x}{4}\right)\ldots\ldots\ldots\infty \text{ terms} = 1 \quad (23)$$

Multiplying both sides of equation (23) and simplifying:

$$\left[\left(1+\frac{x}{1}\right)\left(1-\frac{x}{1}+\frac{x^2}{1}\right)\right]\left[\left(1+\frac{x}{2}\right)\left(1-\frac{x}{2}+\frac{x^2}{2^2}\right)\right]\left[\left(1+\frac{x}{3}\right)\left(1-\frac{x}{3}+\frac{x^2}{3^2}\right)\right]\ldots\ldots =$$

$$\left(1-\frac{x}{1}+\frac{x^2}{1}\right)\left(1-\frac{x}{2}+\frac{x^2}{2^2}\right)\left(1-\frac{x}{3}+\frac{x^2}{3^2}\right)\ldots\ldots \quad (24)$$

Or,

$$\left(1+\frac{x^3}{1^3}\right)\left(1+\frac{x^3}{2^3}\right)\left(1+\frac{x^3}{3^3}\right)\left(1+\frac{x^3}{4^3}\right)\ldots\ldots =$$

$$\beta_1 x + \beta_2 x^2 + \beta_3 x^3 + \ldots\ldots \quad (25)$$

Where β_1, β_2, β_3………..are the coefficients of x, x^2, x^3

respectively in the expansion of right side of the equation (24)

Comparing the coefficients of x^3 on both sides of equation (25), one obtains,

$$1+\frac{1}{2^3}+\frac{1}{3^3}+\frac{1}{4^3}+\ldots\ldots\infty \text{ terms} = |\beta| + \epsilon_2 \qquad (26)$$

Where ϵ_2 is the error caused by the comparison of unaccounted powers of x other than the cubic power on the left and right sides of equation (25)

Subtracting equation (22) from equation (26) and simplifying, one obtains:

$$1+\frac{1}{2^3}+\frac{1}{3^3}+\ldots\ldots\infty \text{ terms} = 4[\beta - \alpha_5] + \frac{3}{4} \qquad (27)$$

By Gauss theorem:
(Error term of boundary condition 1) mod (Error term, boundary condition 2) $\equiv$ 3mod 4 i.e., the factor 3/4 on the right side of equation (27)

Adding equation (22) and equation (26) and simplifying one obtains,

$$1+\frac{1}{2^3}+\frac{1}{3^3}+\ldots\ldots\ldots\ldots\infty\ terms = \frac{5}{8}[\alpha_3+\beta_3]-\frac{1}{4} \qquad (28)$$

In equation (28), -1/4 is obtained by the virtue of Gauss theorem and conforming to our convention.

Now I see with eye serene
The very pulse of machine

— William Wood worth

She was the phantom of delight   *(1804)*

God cannot play dice with the Universe   -Albert Einstein
(1879-1955)

4

The Numeric Wave Equation

Everywhere order reigns, so that when some circumstance has been noted we can foresee that others will be present. The progress of science consist in observing these interconnections and in showing with patient ingenuity that the events of this ever shifting world are what examples of few general connections or relation called laws. To see what is general, what is particular, what is permanent and what is transitory is the aim of scientific thought.

- Alfred North Whitehead (1861-1947)

The numeric wave equation is the fundamental development of replacing complex wave equation e.g. Schrödinger Wave Equation to the ordinary discreet wave type.

Fundamentally this improves the portability and applicability of the equation to more suitable algorithms as well as for the design of the numeric algorithms for complex electronics devices and data communication equipment.

To start with let us consider the sine wave equation in Euler's form viz.

$$\sin(x) = x \prod_{n=1}^{\infty}\left[1 - \frac{x^2}{(n\pi)^2}\right] \qquad (1)$$

But the above equation is designed on premise that $\sin(x)=0$ when $x \neq 0$ and therefore,

$$1 - \frac{(x\pi)^2}{3!} + \frac{(x\pi)^4}{5!} - \frac{(x\pi)^6}{7!} + \ldots \ldots \infty = \prod_{n=1}^{\infty}\left[1 - \frac{x^2}{n^2}\right] = 0 \qquad (2)$$

The equation (1) is 0; for global consideration setting it to 1

$$\prod_{n=1}^{\infty}\left[1 - \frac{x^2}{n^2}\right] = 1 \qquad (3)$$

Setting x=1 in (3) by Riemann Hypothesis or otherwise we have

$$\prod_{n=2}^{\infty}\left[1-\frac{1}{n^2}\right]=\frac{1}{2} \qquad (4)$$

For n=1 the equation (4) takes the value 0 which is set to 1

In fact,

$$\left[1-\frac{1}{n^2}\right]^n \qquad \text{For } n\to\infty \text{ is 1}$$

$$\prod_{n=\infty}^{\infty}\left[1-\frac{1}{n^2}\right]=1$$

Therefore, the entire gamut of roots lie along the axis x=1/2 or of the form

$s=1/2+ i\alpha$

Now the question arises as how they are in error to the unity or ½.

If we take the wave and divide it into N parts say 1,2,3,4.. ..N where N→∞, the value S_n at the element T_n is given by:

$$S_n = \frac{T_n^2}{T_n^2 - 1}\left(\frac{N+1}{N}\right) \text{ for } T_n \neq 1 \qquad (5)$$

$$S_n = \frac{N+1}{N} \text{ for } T_n = 1$$

Similarly, the inverse

$$S_n^{-1} = \frac{T_n^2 - 1}{T_n^2}\left(\frac{N}{N+1}\right) \text{ for } T_n \neq 1 \qquad (6)$$

$$S_n^{-1} = \frac{N}{N+1} \text{ for } T_n = 1$$

Example: Let us consider any wave divided into say N=1000 parts i.e. 1,2,3,4,..........1000 and say we want to find the value at say 50th part, from the formula (5) we have:

$$S_n = \frac{50^2}{50^2 - 1}\left(\frac{1000+1}{1000}\right)$$

For N=1000000 and say we want to find value at 1500th part, again from the formula (5) we have,

$$S_n = \frac{1500^2}{1500^2 - 1}\left(\frac{1000001}{1000000}\right) = 1.000001444$$

This explains why complementary functions are used together as a multiple to correlate with the wave equation and mapped to unity. However the Cartesian coordinate system facilitates the accumulation of points along the axis x=1/2 which is the fundamental basis of Riemann Hypothesis and Prime Number Theorem.

In fact, any mathematical function has the residue at particular point. Refer to Appendix 15.

In numeric wave equation, we are concerned with the laws of nature related to numbers so the residue has vast applications in all facets of Electrical and Computer Engineering.

Things have changed in the past two decades

- Bill Gates (1995)

$E = mc^2$
(*Energy equals mass time the speed of light squared*)

- Albert Einstein (1879-1955)

5

The Prime Number Theorem
The Prime Counting Function
The Error or the Correction Factor

Integers are the creation of God while rest is the work of man

- Leopard Kroenecker

As with Discreet Logarithmic problem, the Gauss conjecture for numbers of prime numbers less than a number is given by

$$Li = \int_2^n \frac{1}{\ln x} dx \qquad (1)$$

The difference between actual numbers of Prime Numbers less than the given number i.e. π(x) and the Gauss conjecture Li(x) is known as error in the Prime Number Theorem.

68 -Riemann Hypothesis and PNT

Let us consider the Numeric Wave Equation (Chapter 4), which precisely state the boundary condition for even powered zetas; i.e. if the wave is divided into N different segments the boundary value of T_N shall be given by:

$$\frac{T_N^2(N+1)}{2(T_N^2-1)N} \qquad (2)$$

On the boundary condition when $T_N = N$ and per convention omitting the factor of ½ (the real part of Riemann Zeta function) we obtain the boundary conditions from equation (2) as:

$$\left(\frac{N}{N-1}\right) \qquad (3a)$$

And

$$\left(\frac{N}{N+1}\right) \qquad (3b)$$

Recalling,

$$\frac{1}{1-x} = 1 + x + x^2 + x^3 + \ldots\ldots$$
and
$$\frac{1}{1+x} = 1 - x + x^2 - x^3 + \ldots\ldots \tag{4}$$

From (4) one obtains:

$$Ln(1-x) = x + \frac{x^2}{2} + \frac{x^3}{3} + \frac{x^4}{4} + \ldots\ldots$$
and
$$Ln(1+x) = x - \frac{x^2}{2} + \frac{x^3}{3} - \frac{x^4}{4} \ldots\ldots\ldots \tag{5}$$

Taking logarithm of 3(a) one obtains:

$$-Ln\left(1-\frac{1}{N}\right) = \frac{1}{N}\left(\sum_{n=2}^{\infty}\frac{1}{n}\right) + \frac{1}{2N^2}\left(\sum_{n=2}^{\infty}\frac{1}{n^2}\right) + \frac{1}{3N^3}\left(\sum_{n=2}^{\infty}\frac{1}{n^3}\right) + \ldots\ldots \tag{6}$$

Performing similar operation on 3(b):

$$-Ln\left(1+\frac{1}{N}\right) = \frac{1}{N}\left(\sum_{n=2}^{\infty}\frac{1}{n}\right) - \frac{1}{2N^2}\left(\sum_{n=2}^{\infty}\frac{1}{n^2}\right) + \frac{1}{3N^3}\left(\sum_{n=2}^{\infty}\frac{1}{n^3}\right) - \ldots\ldots \tag{7}$$

Note that:

$$\sum_{n=2}^{\infty}\frac{1}{n} = \gamma + Ln(n) - 1 = \gamma - 1$$
with Limits $n \xrightarrow{Lim} \infty$, setting $Ln(n) = Ln(\infty) \equiv Ln(1) = 0$

Adding (6) and (7) yields,

$$-\left[Ln\left(1-\frac{1}{N}\right)+Ln\left(1+\frac{1}{N}\right)\right]=\frac{1}{N}(\gamma-1)+\frac{1}{3N^3}\left(\sum_{n=2}^{\infty}\frac{1}{n^3}\right)+\frac{1}{5N^5}\left(\sum_{n=5}^{\infty}\frac{1}{n^5}\right)+\ldots = |\epsilon|$$

.... (8)

Where ϵ is the error factor

As per our convention and Gauss theorem, this error should be 3 mod 4. Since the left side of the equation is negative, the error shall be 1 mod 4.

Therefore the Prime Number less than any given number is

$$\pi(x)=\int_{2}^{N}\frac{1}{Ln(N-\frac{1}{4}\epsilon)}$$

Where ϵ is given by (8)

Chapter 5 The Prime Number Theorem

I shall speak in round numbers, not absolutely accurate yet not so wide from truth so as to vary the result materially

-Thomas Jefferson, (1662)

It seems impossible that anything shall really alter the series of things without the same power which first produced them

-Edward Stillingfleet (1662)

APPENDIX 1

The Quadratic Residue and Euler's Criterion

The modular function of the form

$$x^2 = a \bmod p$$

Where x is an integer and $a \perp p$

For example:
$$23 \bmod 14 = 9 = (\pm 3)^2$$

For a number to be a quadratic residue, the following condition is universal:

$$a^{\frac{p-1}{2}} = 1 \bmod p \qquad (1)$$

In fact the equation (1) bears direct relationship with the Euler's famous results, $\zeta(2n)$ where $n \in \mathbb{N}$

The Least Residue:

The least residue denoted by $L_n M$ is defined as the set of values modulus M i.e.

$$L_n M \equiv \{-(n-1)/2, \ldots\ldots\ldots-3, -2, -1, 0, 1, 2, 3\ldots\ldots\ldots(n-1)/2\}$$

e.g. $L_{21} 30$ would be $30 \bmod 21 \equiv 9$ and $L_{21} 20 = -1$

Convention:

The Legendre Symbol:

$$\left(\frac{a}{p}\right) \equiv \begin{array}{l} 1 \; if \; a \; is \; the \; quadratic \; residue \; \bmod p \\ 0 \; if \; a|p \\ -1 \; otherwise \end{array}$$

The Gauss lemma:

If there are integers m and p such that $m \perp p$ and the least residue is negative, then

$$\left(\frac{m}{p}\right) = (-1)^{\frac{p-1}{2}}$$

Proof: We can represent the set of numbers as

{m, 2m, 3m,................, m(p-1)/2}

In taking modulus the (-) ve sign gets multiplied either even times or odd times and based on the quadratic residue convention and theorem formulation (1), shall be 1 according to whether (p-1)/2 is odd or even.

QED

76 -Riemann Hypothesis and PNT

Appendix 2

The Gauss Theorem

Arithmetic is a certain and infallible art
— Thomas Hobbes
Most numerical analysts have no interest in arithmetic
— B. Parlett (1979)

Gauss Theorem was developed by Gauss when he was seventeen and is fundamental important to the number theory. This theorem baffled the treatises earlier developed by famous mathematicians Legendre and Euler. The theorem established the status of number theory to that of "Queen of Mathematics". In some sense it discerns between number theory and rest of the mathematics. Gauss theorem was so much of fundamental importance that he returned to it many times in his later life and gave many different proofs. The following proof by author is predicted to be the 153rd proof.

Theorem:

Let p and q be distinct prime, then

$$\left(\frac{p}{q}\right) = \left(\frac{q}{p}\right)$$

If $p \equiv q \equiv 3 \mod 4$ in which case

$$\left(\frac{p}{q}\right) \neq \left(\frac{q}{p}\right)$$

In other words

$$\left(\frac{p}{q}\right) \equiv \left(\frac{q}{p}\right) \equiv (-1)^{\frac{(p-1)(q-1)}{4}}$$

E.g. if p=11 and q=19 establishes that

$$\left(\frac{p}{q}\right) \neq \left(\frac{q}{p}\right)$$

Appendix 2 The Gauss Theorem 79

It can be easily checked that for p= 11 and q=19, $p \equiv q \equiv 3$ mod 4

Now consider p= 23 and q=53, here $p \equiv 3$ mod 4 but $q \neq 3$ mod 4 and the product (p-1)(q-1)/4 = 256 an even number, therefore

$$\left(\frac{p}{q}\right) = \left(\frac{q}{p}\right)$$

This reciprocity theorem is the extension of quadratic residue and shows how marvelously the prime numbers internally and elegantly takes care of positive and negative signs in the reciprocity. To prove the above theorem let us consider that the prime number p runs like

{p, 2p, 3p,...............{(q-1)/2}p

And the prime number q runs like

{q, 2q, 3q,...............{(p-1)/2}q

Proof 1:

From Chinese Remainder Theorem we have a unique number modulo product of primes like the pigeon hole. Similarly Euler's product formula satisfies the solution for cyclicity. The reduced form of Euler's product formula can be represented as the product of two functions viz.

$$\sin x = x \prod_{n=1}^{\infty}\left[1 - \frac{x^2}{(n\pi)^2}\right]$$

$$\prod_{n=1}^{\infty}\left[1 - \frac{x^2}{(n\pi)^2}\right] = F(x).F(-x) \qquad (1)$$

The boundary value of $F(x).F(-x)$ is obvious by considering negative least residue 1 mod 4 $\equiv$ 3 mod 4 for odd zetas i.e. $\zeta(3); \zeta(5); \zeta(7);\ldots\ldots$

Appendix 2 The Gauss Theorem 81

Proof 2: Proof by geometric considerations

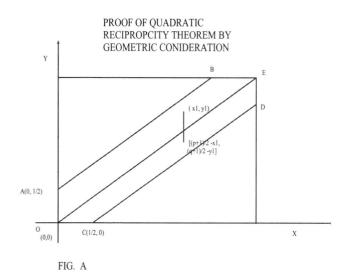

FIG. A

According to the definition, the number of integers multiple of prime number $p = (q-1)/2 = \alpha$ and the number of integers multiple of prime number $q = (p-1)/2 = \beta$

The Gauss Lemma:

$\left(\dfrac{p}{q}\right) = \left(\dfrac{q}{p}\right)$ if and only if

$(-1)^{\alpha+\beta} = 1$

To prove quadratic reciprocity theorem we need to prove that:

$(-1)^{\alpha+\beta} = -1$ or $\alpha + \beta$ is odd

If and only if p $\equiv$ q $\equiv$ 3 mod 4

Considering Fig. A, the equation of line AB is y= (p/q)x + ½

Similarly, the equation of line CD is x = (q/p)y +1/2. Lines AB, CD (and OE) are parallel.

Let x_1 and y_1 be the coordinates inside the hexagon. By the boundary equations it can be shown that point (m, n)$\equiv$[(p +1)/2 - x_1], [(q +1)/2 - y_1] lies within the OABCDE.

The only way the point $(x_1, y_1) \equiv$ (m, n) is m: (p+1)/4 and n: (q+1)/4 which is possible only when p$\equiv$q$\equiv$ 3 mod 4 for distinct primes p and q.

QED

Appendix 2 The Gauss Theorem

There is no sect in geometry.

> \- Voltaire (1694-1778)
> Philosophical Dictionary

Who knows not geometry, not enter here.

> \- Anonymous (Greek)
> Inscription at the entrance of Plato's home

The most beautiful emotion we can experience is mystical. It is the power of all true art and science.

> \- Albert Einstein (1879-1955)

Appendix 3

The Properties of Zeta Function $\zeta(I)$

The science of pure mathematics, in its modern development may claim to be most original creations of human spirit.
— Alfred Noble Whitehead (1861-1947)

The properties of $\zeta(I)$ where I is the positive integers is important because of the internal working of computer depends upon the integer functionality. Computers can work more precisely on integers than on the floating point computations. In fact, the floating computations are just approximation.

$$\sum_{I=2}^{\infty}[\zeta(I)-1]=1 \qquad (1)$$

$$\sum_{I=2}^{\infty}[\zeta(2I-1)-1]=\frac{1}{4} \qquad (2)$$

$$\sum_{I=1}^{\infty}[\zeta(2I)-1]=\frac{3}{4} \qquad (3)$$

$$\sum_{I=2}^{\infty}(-1)^{I}[\zeta(I)-1]=\frac{1}{2} \qquad (4)$$

Proof:

$$\zeta(2)-1=\frac{1}{2^2}+\frac{1}{3^2}+\frac{1}{4^2}+\ldots\ldots$$

$$\zeta(3)-1=\frac{1}{2^3}+\frac{1}{3^3}+\frac{1}{4^3}+\ldots\ldots$$

$$\vdots$$

$$\zeta(\infty)-1=\frac{1}{2^{\infty}}+\frac{1}{3^{\infty}}+\frac{1}{4^{\infty}}+\ldots\ldots$$

Adding the above equations

$$=\frac{1}{1.2}+\frac{1}{2.3}+\frac{1}{3.4}+\frac{1}{4.5}+\ldots\ldots\infty=1$$

Similarly,

$$\sum_{I=2}^{\infty}[\zeta(2I-1)-1]=\left[\frac{1}{2^3}\left(1-\frac{1}{2^2}\right)^{-1}\right]+1\left[\frac{1}{3^3}\left(1-\frac{1}{3^2}\right)\right]+\ldots\ldots$$

$$= \frac{1}{2.3} + \frac{1}{3.8} + \frac{1}{4.15} + \frac{1}{5.24} + \ldots \ldots \infty = \frac{1}{4}$$

For (3) one obtains:

$$\left[\frac{1}{2^2} + \frac{1}{2^4} + \frac{1}{2^6} + \ldots \ldots \right] + \left[\frac{1}{3^2} + \frac{1}{3^4} + \frac{1}{3^6} + \ldots \ldots \right] + \ldots \ldots \ldots \infty$$

$$= \frac{1}{2^2 - 1} + \frac{1}{3^2 - 1} + \frac{1}{4^2 - 1} + \ldots \ldots \infty = \frac{1}{1.3} + \frac{1}{2.4} + \frac{1}{3.5} + \frac{1}{4.6} + \ldots \ldots \infty = \frac{3}{4}$$

From (2) and (3) one obtains the relation (4)

Now let's consider another important relation of zeta function:

$$\sum_{i=2}^{\infty} \frac{1}{i} [\zeta(i) - 1] = 1 - \gamma \qquad (5)$$

Proof:
By definition,

$$\gamma = \sum_{k=1}^{\infty} \frac{1}{k} - Lnk = 1 + \sum_{k=2}^{\infty} \frac{1}{k} - \sum_{k=2}^{\infty} Ln\frac{k}{k-1} = 1 + \sum_{k=2}^{\infty} \frac{1}{k} + \sum_{k=2}^{\infty} Ln\frac{k-1}{k} \qquad (6)$$

$$= 1 + \sum_{k=2}^{\infty} \left[\frac{1}{k} + Ln\left(1 - \frac{1}{k}\right) \right] = 1 - \sum_{k=2}^{\infty} \left[\sum_{l=2}^{\infty} \frac{1}{lk^l} \right]$$

$$= 1 - \sum_{l=2}^{\infty} \frac{1}{l} \left[\zeta(l) - 1 \right]$$

The equation (6) can be calculated to any digit on the large processors.

Or, $\gamma \approx 0.57721\ 56649\ 01532\ 86060\ 6512\ldots\ldots\ldots$ γ is called Euler's constant

$$= n \xrightarrow{Lim} \infty \left[1 + \tfrac{1}{2} + \tfrac{1}{3} + \tfrac{1}{4} + \ldots .. \tfrac{1}{n} - \ln n \right]$$

In this appendix we note that zeta function play a very symmetric role in the Cartesian coordinate system. These properties are the foundation of finding the subtle but definite deviation in residue of the functions at the particular point by real rather than imaginary or complex numbers. Most importantly it let us find the error or the correction factor of the Prime Number Theorem.

Appendix 4
Fermat's Little Theorem and Chinese Remainder Theorem

The scholar who is not grave will not inspire respect and his learning will therefore lack stability. His chief principles must be conscientiousness and sincerity. Let him have no friend unequal to him. And let him not hesitate to amend himself when in wrong.

- Confucius

In reference to the Gauss theorem, we shall mention two important theorems:

(1) Fermat's Little Theorem: If p is prime and n is a positive integer then

$$p \mid n^{p-1} - 1$$

For a special case if g.c.d.(n,p)=1

$$p \mid n^{p-1} - 1$$

(2) Wilson's Theorem: If p is prime number then

$$p\,|\,[(p-1)!+1]$$

Another important theorem is Remainder theorem studied by Chinese Mathematician Sun-Tzu which is as follows:

Let $m_1, m_2, m_3 \ldots\ldots m_p$ are p integers such that there is no common factor between them other than 1. Or, $g.c.d.(a_j, m_j) = 1$ and $M = m_1 m_2 m_3 \ldots\ldots m_p$

Suppose there are integers $a_1, a_2, a_3 \ldots\ldots\ldots a_p$

The p congruence

$$a_1 x \equiv b_1 \pmod{m_1}$$

$a_2 x \equiv b_2 \pmod{m_2}$

............,

$a_p x \equiv b_p \pmod{m_p}$

have a simultaneous solution that is unique modulo M.

The Chinese Remainder Theorem has major applications in

App. 4 Fermat's Th./ Chinese Remainder Th.-91

Computer Science and Engineering.

Example of Chinese Remainder Theorem:

Find a number which when divided by 3 leaves the remainder 2, when divided by 5 leaves the remainder 4 and when divided by 7 leaves the remainder 6.

We note that 3, 5, and 7 are prime (or mutually) prime numbers.

Or, M = 3x5x7 = 105

Now a_1 = 105/3 = 35, a_2 = 105/5 = 21 and a_3 = 105/7 = 15

35.x = 3 mod 105
21.x = 5 mod 105
15.x = 7 mod 105

Reducing these equations yields:

2.x = 3 mod 105
1.x = 5 mod 105
1.x = 7mod 105

Or, the unique solution would be ((number N) mod 105)

Where N = 2x2x35 + 1x4x21 + 1x6x15 = 140 + 84 + 90 = 314

Or the unique solution = 314 mod 105 = 104

APPENDIX 5
Special Functions
For
Prime Number Theorem

From the intrinsic evidence of his creation, the Great Architect Design of the Universe, now begins to appear in pure mathematics.

-James Hopwood Jeans

The real world applications of mathematics are non-linear type. By the virtue of converging infinite series the real world applications are simulated. Here are some formulas and engineering methods to calculate otherwise difficult functions for their inexpressibility in conventional mathematics:

(1) The error function:

$$erf(x) = \frac{2}{\sqrt{\pi}} \int_0^x e^{-u^2} du$$

$$erf(x) = \frac{2}{\sqrt{\pi}}(x - \frac{x^3}{3.1!} + \frac{x^5}{5.2!} - \frac{x^7}{7.3!} + \ldots)$$

$$erf(-x) = -erf(x), \quad erf(0) = 0, \quad erf(\infty) = 1$$

(2) The Exponential Integral:

$$E(x) = \int_x^\infty \frac{e^{-u}}{u} du$$

$$E(x) = -\gamma - \ln x + \int_0^x \frac{1 - e^{-u}}{u} du$$

$$E(x) = -\gamma - \ln x + \left(\frac{x}{1.1!} - \frac{x^2}{2.2!} + \frac{x^3}{3.3!} - \ldots\right)$$

$$E(\infty) = 0$$

(3) The Sine Integral

Appendix 5 Special Function for PNT-95

$$Si(x) = \int_0^x \frac{\sin u}{u} du$$

$$Si(x) = \frac{x}{1.1!} - \frac{x^3}{3.3!} + \frac{x^5}{5.5!} - \frac{x^7}{7.7!} + \dots$$

$$Or, \ Si(x) = \sum_{r=1}^{\infty} (-1)^{r-1} \frac{x^{2r-1}}{(2r-1)(2r-1)!}$$

$$Si(x) \approx \frac{\pi}{2} - \frac{\sin x}{x}\left(\frac{1}{x} - \frac{3!}{x^3} + \frac{5!}{x^5} - \dots\right) - \frac{\cos x}{x}\left(1 - \frac{2!}{x^2} + \frac{4!}{x^4} - \dots\right)$$

$$Si(-x) = -Si(x), \quad Si(0) = 0, \quad Si(\infty) = \frac{\pi}{2}$$

(4) The Cosine Integral

$$Ci(x) = \int_x^{\infty} \frac{\cos u}{u} du$$

$$Ci(x) = -\gamma - \ln x + \int_0^x \frac{1 - \cos u}{u} du$$

$$Ci(x) = -\gamma - \ln x + \frac{x^2}{2.2!} - \frac{x^4}{4.4!} + \frac{x^6}{6.6!} - \frac{x^8}{8.8!} + \dots$$

$$Or, \ Ci(x) = -\gamma - \ln x - \sum_{r=1}^{\infty} \frac{(-x^2)^r}{2r(2r)!}$$

$$Ci(x) \approx \frac{\cos x}{x}\left(\frac{1}{x} - \frac{3!}{x^3} + \frac{5!}{x^5} - \dots\right) - \frac{\sin x}{x}\left(1 - \frac{2!}{x^2} + \frac{4!}{x^4} - \dots\right)$$

$$Ci(\infty) = 0$$

(5) The Logarithmic Integral

The logarithmic integral is defined by:

$$Li(x) = \int_2^x \frac{du}{\ln u}$$

For convenience it is acceptable to replace the Li(x) by li(x) i.e.:

$$li(x) = \underset{\eta \longrightarrow 0}{Lim} \left(\int_0^{1-\eta} + \int_{1+\eta}^x \right) \frac{du}{\ln u}$$

Which it differs only by a constant li(2) = 1.04.......

$$li(2) = \gamma + \ln \ln(2) + \sum_{r=1}^{\infty} \frac{[\ln 2]^r}{r \cdot r!}$$

li(2) = 1.045163780117492784844588881946131365226155781512015758329....

Johann Von Soldner (1766-1833) and Srinivas Ramanujan (1887-1920) defined constant μ:

μ = 1.451369234883381050283968485892027449493+

"μ" is the only positive root of logarithmic integral function

Appendix 5 Special Function for PNT-97

'li' which can be considered as

$$Li(x) = li(x) - li(2)$$

$$li(x) = \int_0^x \frac{dt}{\ln t} = \int_\mu^x \frac{dt}{\ln t}$$

Let $\pi(x)$ be the number of primes $\leq x$, such that $\pi(2)=1$, $\pi(10)=4$, $\pi(1000)=168$; the asymptotic behavior of this function has been studied by the world's greatest mathematicians beginning with Legendre in 1798. Numerous advances made during 19th century culminated when Charles de La Vallèe Poussin proved that for some A>0,

$$\pi(x) = \int_2^x \frac{dt}{\ln t} + O\left(xe^{-A\sqrt{\log x}}\right) \qquad (I)$$

Integrating by parts for Li(x) and li(x) yields:

$$li(x) = \frac{x}{\ln x} + \frac{x}{(\ln x)^2} + \frac{2!x}{(\ln x)^3} + \ldots\ldots + \frac{r!x}{(\ln x)^{r+1}} + \ldots C \qquad (II)$$

In fact the relation (I) diverges very fast after a few steps. For r, x→∞, the relation (II) can be mapped to:

98 -Riemann Hypothesis and PNT

$$li(x) = \gamma + \ln \ln x + \sum_{r=1}^{\infty} \frac{(\ln r)^r}{r \cdot r!}$$

$$\pi(x) = \frac{x}{\ln x} + \frac{x}{(\ln x)^2} + \frac{2!x}{(\ln x)^3} + \ldots\ldots + \frac{r!x}{(\ln x)^{r+1}} + O\left(\frac{x}{(\log x)^{r+2}}\right)$$

For all fix $r \geq 0$

Where the error term was:

$$O\left(x \exp\left(-A(\log x)^{3/5} / (\log \log x)^{1/5}\right)\right)$$

B. Riemann conjectured in 1859 that:

$$\pi(x) = \sum_{k=1}^{\lg x} \frac{\mu(k)}{k} L\left(\sqrt[k]{x}\right) + O(1) = L(x) - \tfrac{1}{2} L\left(\sqrt{x}\right) - \tfrac{1}{3} L\left(\sqrt[3]{x}\right) + \ldots\ldots + O(1)$$

Where

$$L(x) = \int_2^x \frac{dt}{\ln t}$$

Eventually Littlewood in 1914 dismissed the Riemann's claim for the formula and stated that the prime counting function requires deep mathematical properties to be established.

Appendix 5 Special Function for PNT-99

The reader may refer to Chapter 5 of this book to be familiar with deep mathematical properties for the Prime Counting Function and Prime Number Theorem.

To evaluate Li(x) there is another formula of S. Ramanujan which converges more rapidly viz.:

$$\int_{\mu}^{x}\frac{dt}{\ln t} = \gamma + \ln \ln x + \sqrt{x}\sum_{n=0}^{\infty}\frac{(-1)^{n-1}(\ln x)^n}{n!2^{n-1}}\sum_{k=0}^{\left[\frac{n-1}{2}\right]}\frac{1}{2k+1}$$

Some important relations in respect to logarithmic function are:

$$\frac{1}{\ln(1+x)} = \frac{1}{x} + \frac{1}{2} - \frac{1}{12} + \frac{1}{24}x^2 - \frac{19}{720}x^3 + \frac{3}{160}x^4 + \dots\dots$$

$$\sum_{k=1}^{\infty}(-1)^k \ln k = \frac{1}{2}\ln\left(\tfrac{1}{2\pi}\right)$$

$$\sum_{k=1}^{\infty}\ln k = \frac{1}{2}\ln(2\pi)$$

$\ln 1 = 0$

$\ln 0 = -\infty$

$\ln e = 1$ such that

100-Riemann Hypothesis and PNT

$$e = n \xrightarrow{Lim} \infty \left(1+\frac{1}{n}\right)^n = 2.7182818284.......$$

$$\text{and} \int_1^e \frac{dt}{t} = 1$$

$\ln(-1) = \pi i$

$\ln(\pm i) = \pm \tfrac{1}{2}\pi i$

Integration formula:

$$\int \frac{dz}{z} = \ln z$$

$$\int \ln z\, dz = z \ln z - z$$

$$\int_0^1 \frac{\ln t}{1-t} = -\frac{\pi^2}{6}$$

$$\int_0^1 \frac{\ln t}{1+t} = -\frac{\pi^2}{12}$$

$$\int_0^x \frac{dt}{\ln t} = li(x)$$

Series Expansion

Appendix 5 Special Function for PNT-101

$$\ln\left(\frac{z-1}{z+1}\right) = 2\left(\frac{1}{z} + \frac{1}{3z^3} + \frac{1}{5z^5} + \ldots\ldots\right)$$

$$(|z| \geq 1, z \neq \pm 1)$$

$$\ln(z+a) = \ln a + 2\left[\left(\frac{z}{2a+z}\right) + \frac{1}{3}\left(\frac{z}{2a+z}\right)^3 + \frac{1}{5}\left(\frac{z}{2a+z}\right)^5 + \ldots\ldots\right]$$

$$(a > 0, \text{Re}\, z \geq -a \neq 0)$$

$$\ln z = 2\left[\left(\frac{z-1}{z+1}\right) + \frac{1}{3}\left(\frac{z-1}{z+1}\right)^3 + \frac{1}{5}\left(\frac{z-1}{z+1}\right)^5 + \ldots\ldots\right]$$

$$(\text{Re}\, z \geq 0, z \neq 0)$$

$$\ln z = (z-1) - \tfrac{1}{2}(z-1)^2 + \tfrac{1}{3}(z-1)^3 - \ldots\ldots$$

$$(|z-1| \leq 1 \text{ and } z \neq -1)$$

$$\ln z = \left(\frac{z-1}{z}\right) + \frac{1}{2}\left(\frac{z-1}{z}\right)^2 + \frac{1}{3}\left(\frac{z-1}{z}\right)^3 + \ldots\ldots$$

$$(\text{Re}\, z \geq \tfrac{1}{2})$$

$$\ln(1+z) = z - \frac{1}{2}z^2 + \frac{1}{3}z^3 - \ldots\ldots$$

$$(|z| \leq 1 \text{ and } z \neq -1)$$

102-Riemann Hypothesis and PNT

APPENDIX 6

Values of the First Few Zeta Functions

Man of science has learned to believe in justification, not by faith but verifications

— T.H. Huxley

The following are some values for the zeta function;

$$\zeta(s) = \sum_{k=1}^{k=\infty} \frac{1}{k^s}$$

$\zeta(1) = \infty$

$\zeta(2) = \dfrac{\pi^2}{6}$

$\zeta(3) = 1.202056903159594\ldots$

$\zeta(4) = \dfrac{\pi^4}{90}$

104-Riemann Hypothesis and PNT

$\zeta(5) = 1.0369277551433............$

$\zeta(6) = \dfrac{\pi^6}{945}$

$\zeta(7) = 1.0083492773819$

$\zeta(8) = \dfrac{\pi^8}{9450}$

$\zeta(9) = 1.002008392826............$

$\zeta(10) = \dfrac{\pi^{10}}{93555}$

$\zeta(12) = \dfrac{691\pi^{12}}{638512875}$

$\zeta(14) = \dfrac{2\pi^{14}}{18243225}$

App. 6 The Value of Few Zeta Functions

$$\zeta(16) = \frac{3617\pi^{16}}{325641566250}$$

$$\zeta(18) = \frac{43867\pi^{18}}{38979295480125}$$

$$\zeta(20) = \frac{174611\pi^{20}}{1531329465290625}$$

$$\zeta(22) = \frac{155366\pi^{22}}{13447856940643125}$$

$$\zeta(24) = \frac{236364091\pi^{24}}{201919571963756218750}$$

$$\zeta(26) = \frac{1315862\pi^{26}}{11094481976030578125}$$

$$\zeta(28) = \frac{6785560294\pi^{28}}{564653660170076273671875}$$

$$\zeta(30) = \frac{6892673020804\pi^{30}}{5660878804669082674070015625}$$

$$\zeta(32) = \frac{7709321041217\pi^{32}}{62490220571022341207266406250}$$

106-Riemann Hypothesis and PNT

APPENDIX 7

The Complex Number

Stand firm in your refusal to remain conscious during algebra. In real life there is no such thing as algebra.
- Fran Libowitz (1951-)

Integers are creation of God, rest all is the work of man.
- Leopard Kroenecker

The complex numbers are the mysterious phase of mathematics and number theory in particular.

The complex number as defined by:

$i = \sqrt{-1}$

It has mysterious properties; e.g. Euler in 1761 found that

108-Riemann Hypothesis and PNT

$$e^{i\theta}+1 = 0$$

$$e^{i\theta} = \cos\theta + i\sin\theta \qquad (1)$$

Our main concern is as what does the equation means. In fact, it does not mean anything but only that we have derived it by logical steps process.

The concept Euler used was is as:

By binomial theorem we have,

$$e = 1 + 1 + \frac{1}{2!} + \frac{1}{3!} + \ldots \ldots \infty$$

$$e^x = 1 + x + \frac{x^2}{2!} + \frac{x^3}{3!} + \frac{x^4}{4!} + \ldots \ldots \infty$$

Putting $x = i\theta$, we obtain (1)

We can derive cosh (x) and sh (x) in terms e. In the similar fashion cos(x) and sine(x) are represented in form of and as:

$$sh(x) = \frac{e^x - e^{-x}}{2}$$

Appendix 7 The Complex Number-109

$$\cosh(x) = \frac{e^x + e^{-x}}{2}$$

Here x can be just any arbitrary number. Sh (x) and Cosh (x) are hyperbolic function formed by 2xy rotating the coordinate axis by in counterclockwise direction but still only matches asymptotic in relation to the new virtual axis. Similarly,

$$\sin(x) = \frac{e^{ix} - e^{-ix}}{2i} \qquad \cos(x) = \frac{e^{ix} + e^{-ix}}{2}$$

$$\sin^2 x + \cos^2 x = 1$$

is a general identity of sine and cosine relationship and bears the exact relation to the equation

$$x^2 + y^2 = 1$$

Where the radius is unity and (x, y) are coordinates along the circumference of the circle with center in origin (0, 0) of the coordinate system. But

$$sh^2(x) - \cosh^2(x) = 1$$

110-Riemann Hypothesis and PNT

When the two hyperbolic functions are the same? Are they same at the limit ∞ and limit - ∞?

A perfunctory look at the hyperbolic sine and cosine functions exhibit that at limit ∞ they satisfy the asymptotic behavior while at the limit - ∞ they do not satisfy. This can be explained in terms of simple behavior viz.:

$$Log_e 1 = Log_e(-1)^2 = 2 Log_e(-1) = 2^n Log_e(-1)$$

Where n∈N

We can simply put this concept in other way viz.:

The derivative of e^x is $\frac{d(e^x)}{dx} = e^x$

Or, in other words integral of e^x is

$$\int e^x dx = e^x + C$$

Where C is Constant of Integration

By definition

$$e^x = n \to \infty \left(1 + \frac{x}{n}\right)^n = 1 + \frac{x}{1!} + \frac{x^2}{2!} + \frac{x^3}{3!} + \ldots \ldots \infty \qquad (2)$$

Or,

$$e^{-x} = n \to \infty \left(1 - \frac{x}{n}\right)^n = 1 - \frac{x}{1!} + \frac{x^2}{2!} - \frac{x^3}{3!} + \ldots \ldots \infty \qquad (3)$$

Multiplying (2) and (3) yields

For $n \to \infty$

$$\left(1 - \frac{x^2}{n^2}\right)^n = 1 \qquad (4)$$

Putting x=1 in equation (4) we obtain

$$\left(1 - \frac{1}{n^2}\right)^n = 1 \text{ for } n \to \infty$$

In fact the equation (4) is of the form

$$\frac{1}{2}\prod_{2}^{\infty}\left(1 - \frac{1}{p^2}\right)$$

Where $p \in \mathbb{N}$, i.e. 1, 2, 3, 4...∞ which is Euler's sine product formula and the fundamental result for the Numeric Wave Equation. For details of Numeric Wave Equation, please refer to the Chapter 4.

The beauty of the Numeric Wave Equation is that it eliminates the need for iota $i=\sqrt{-1}$ and thus we can objectively deal with the concept of complex number for real world applications.

APPENDIX 8
Riemann Zeta Function Analysis I

If the scientist uncovers a publishable fact, it will become central to his theory.

— Albert Bloch

"Advanced Research-man-ship"

Riemann was trained in complex analysis; in fact he was one of the founders of complex theory. In addition, he was trained under G.F. Gauss and Dirichlet and was influenced by Chebyshev and Legendre. Yet the original basis of the zeta function was developed by L. Euler, when he discovered the series

$$1 + \frac{1}{2^2} + \frac{1}{3^2} + \ldots\ldots\ldots\infty \text{ terms} = \frac{\pi^2}{6}$$

The method used by L. Euler was simple algebraic as follows:

$$\sin x = x - \frac{x^3}{3!} + \frac{x^5}{5!} - \frac{x^7}{7!} \ldots\ldots\infty \qquad (1)$$

114-Riemann Hypothesis and PNT

When

$\sin x = 0 \cap x \neq 0$; (1) becomes,

$$\sin x = 1 - \frac{x^2}{3!} + \frac{x^4}{5!} - \frac{x^6}{7!} + \ldots\ldots\infty = 0 \qquad (2)$$

Therefore the roots of (2) are, namely

$\pm \pi, \pm 2\pi, \pm 3\pi, \pm 4\pi, \ldots\ldots\infty$

Using the Factorial or the Remainder Theorem we have,

$$\left(1 - \frac{x^2}{(\pi)^2}\right)\left(1 - \frac{x^2}{(2\pi)^2}\right)\left(1 - \frac{x^2}{(3\pi)^2}\right)\left(1 - \frac{x^2}{(4\pi)^2}\right)\ldots\ldots\infty = 0 \qquad (3)$$

Comparing the coefficients of in equations (2) and (3), we obtain the Euler's famous result.

The equation (3) clearly gives the product to even powers of x. Euler carried his algebraic computations up to the 26th powers viz.

$$\zeta(26) = 1 + \frac{1}{2^{26}} + \frac{1}{3^{26}} + \frac{1}{4^{26}} + \ldots\ldots\infty = \frac{1315862\pi^{26}}{11094481976030578125}$$

(4)

App 8 Riemann Zeta Function- Analysis 1-115

In fact the equation (4) can be extended up to infinity **even** powers i.e.:

$$\zeta(\infty) = 1 + \frac{1}{2^\infty} + \frac{1}{3^\infty} + \frac{1}{4^\infty} + \ldots\ldots\ldots\ldots\infty = \frac{p}{q}\pi^\infty$$

Where p, q $\in \mathbb{N}$

Euler deduced,

$$1 + \frac{1}{2^4} + \frac{1}{3^4} + \frac{1}{4^4} + \ldots\ldots\ldots\infty = \frac{\pi^4}{90}$$

On the same basis he discovered,

$$\zeta(6) = 1 + \frac{1}{2^6} + \frac{1}{3^6} + \frac{1}{4^6} + \ldots\ldots\ldots\infty = \frac{\pi^6}{945}$$

$$\zeta(8) = 1 + \frac{1}{2^8} + \frac{1}{3^8} + \frac{1}{4^8} + \ldots\ldots\ldots\infty = \frac{\pi^8}{9450}$$

And so on...

Subsequently, L. Euler formulated the general solution for

116-Riemann Hypothesis and PNT

even zetas viz.:

$$\zeta(2n) = \sum_{k=1}^{\infty} \frac{1}{k^{2n}} = (-1)^{n-1} \frac{(2\pi)^{2n}}{2(2n)!} B_{2n}$$

Where B_{2n} are the Bernoulli Numbers

The above series are convergent series. If the power happens to be 1 the series diverges, i.e.

$$1 + \frac{1}{2} + \frac{1}{3} + \frac{1}{4} + \ldots \ldots \infty \text{ terms} = \infty \qquad (5)$$

The series (5) is called harmonic series and it diverges. Typically, for s > 1, the ζ(s) converges and it diverges for s≤1

$$\zeta(s) = 1 + \frac{1}{2^s} + \frac{1}{3^s} + \frac{1}{4^s} + \ldots \ldots \infty$$

What's the relationship between zeta function and the prime numbers?

Let's consider a zeta function viz.;

App 8 Riemann Zeta Function- Analysis 1-117

$$\zeta(s) = 1 + \frac{1}{2^s} + \frac{1}{3^s} + \frac{1}{4^s} + \frac{1}{5^s} + \frac{1}{6^s} + \frac{1}{7^s} + \ldots \infty \qquad (6)$$

Multiplying both sides of equation (6) by $\frac{1}{2^s}$ and subtracting from equation (6) one obtains:

$$\left(1 - \frac{1}{2^s}\right)\zeta(s) = 1 + \frac{1}{3^s} + \frac{1}{5^s} + \frac{1}{7^s} + \frac{1}{9^s} + \frac{1}{11^s} + \frac{1}{13^s} \ldots \infty \qquad (7)$$

Proceeding in the same way and simplifying,

$$\sum_n \frac{1}{n^s} = \prod_p \frac{1}{\left(1 - \frac{1}{p^s}\right)}$$

$$\zeta(s) = \sum_n n^{-s} = \prod_p (1 - p^{-s})^{-1}$$

Here the n ranges for all positive integers, i.e. n=1,2,3,4.........∞ and p ranges for all prime numbers i.e. p=2,3,5,7..........∞.

$$\zeta(3) = \prod_{p=prime} (1 - p^{-3}) = 1.20205\ldots\ldots$$

Euler on similar basis was able to prove:

118-Riemann Hypothesis and PNT

$$1 - \frac{1}{3^3} + \frac{1}{5^3} - \frac{1}{7^3} + \frac{1}{9^3} - \frac{1}{11^3} + \ldots \ldots \infty = \frac{\pi^3}{32}$$

$$1 - \frac{1}{3^5} + \frac{1}{5^5} - \frac{1}{7^5} + \frac{1}{9^5} - \frac{1}{11^5} + \ldots \ldots \infty = \frac{5\pi^5}{1536}$$

Or in most general form,

$$1 - \frac{1}{3^{(2n+1)}} + \frac{1}{5^{(2n+1)}} - \frac{1}{7^{(2n+1)}} + \ldots \ldots \infty = \frac{p}{q} \pi^{(2n+1)} \tag{8}$$

In fact it can be easily shown that

$$1 - \frac{1}{3^{(2n+1)}} + \frac{1}{7^{(2n+1)}} - \frac{1}{13^{(2n+1)}} + \frac{1}{17^{(2n+1)}} - \ldots \ldots \infty = \frac{p_1}{q_1} \pi^{(2n+1)} \tag{9}$$

$$1 - \frac{1}{5^{(2n+1)}} + \frac{1}{9^{(2n+1)}} - \frac{1}{19^{(2n+1)}} + \frac{1}{23^{(2n+1)}} - \ldots \ldots \infty = \frac{p_2}{q_2} \pi^{(2n+1)} \tag{10}$$

And so on.

Where p, q, p_1, q_1... and n $\in \mathbb{N}$

It may be easily observed that Bernoulli's numbers are

related to even zeta function $\zeta(2n)$; the series of type (8) has its roots in Euler's number while other series such as (9) or (10) have no numbers assigned to them. Therefore we apparently assign the numbers as Euler's extension number (Refer to Appendix 15)

Dirichlet extended the concept of $\zeta(s)$ to all the real powers for s>1 and gave rigorous proof. Riemann extended the concept to any complex numbers. The function Riemann defined was meaningful for all values of s with the exception at pole s =1. This extension is supported through a function called a factorial function or gamma function.

The Gamma Function:
Gamma Function defined originally as Factorial function for Re s >0 is given by:

$$\Gamma(s) = \int_0^\infty e^{-x} x^{s-1} dx \qquad (11)$$

And for all other value of s by analytical continuation s a regular function with poles at s=0,-1,-2,-3........∞

120-Riemann Hypothesis and PNT

The Gamma Function or the Factorial Function has its roots in Euler's extension of the factorial function from natural number to real number. For natural number,

n! = 1.2.3.4..........(n-2).(n-1).n.

Euler observed that

$$n! = \int_0^\infty e^{-x} x^n dx$$

For n∈ℕ

Gauss observed that the right side of the integral converges for n>-1 and introduced a notation:

$$\Pi(s) = \int_0^\infty e^{-x} x^n dx \text{ for } s > -1 \tag{12}$$

The expression (11) is good for all real or rather all complex values in half plane s >-1

Prior to Legendre, L. Euler formulated expression (12). Legendre considered that half plane to start at 0 hence preferred definition (11) over (12)

$\Gamma(1) = 1 = 0!$

App 8 Riemann Zeta Function- Analysis 1-121

$\Gamma(s+1) = s\Gamma(s)$

If s is positive integer, from (12) we have,

$\Gamma(s+1) = s(s-1)(s-2)......3.2.1 = s!$

If s =½ we have

$\Gamma(½) = \sqrt{\pi} = \left(-\dfrac{1}{2}\right)!$

There are infinitely many values can be generated this way. For Re(s)≤0 the definition (11) lapses since this integral diverges. By analytical continuation we can define in the left hand plane. This is equivalent to switching to (12) for which the gamma function has simple poles at s = 0,-1,-2...∞
It can be easily shown

$\Gamma(x)\Gamma(x-1) = \dfrac{\pi}{\sin(\pi x)}$ (12a)

$\zeta(x)\Gamma(x) = \displaystyle\int_0^\infty \dfrac{u^{x-1}}{e^u - 1} du$ (12b)

The gamma function was extensively used by Riemann and

in mathematics literature has it's weakness for not able to satisfy the quadratic reciprocity and as per equation (12a), it is equivalent to Euler's sine product formula via its complementary multiplication i.e. $\Gamma(x).\Gamma(1-x)$. It misses out the Gauss Reciprocity theorem for the Prime Numbers and hence the deep properties of Prime Number Theorem or the Discreet Logarithmic problem. Or, in other words it doesn't directly felicitate the proof of closing value of Apery Constant and hence the deep properties surrounding prime numbers. The functional equation of Riemann Zeta function presents this reference with the trivial zeros at -2, -4, -6,.... However, gamma function is closely related to $\gamma= 0.577...$ and e=2.71... for it's unilateral (positive infinite product) relationship. But gamma function does not truly represent the laws of nature governing numbers. This is because the gamma function is based on only positive infinite product and it misses out inner and subtle effect of the boundary conditions required for negative infinite product analysis for satisfying the distribution the prime numbers.

Appendix 9

The Riemann Zeta Function Analysis 2

No great discovery was ever made in science except by one who lifted his nose above the grindstone of details and ventured on a more comprehensive vision.

-Albert Einstein(1879-1955)

Bernhard Riemann's eight page paper entitled " On the Number of Prime Less than a Given Magnitude" was a landmark feet and conjecture that influenced generations of great mathematicians such as S. Ramanujan, G.H. Hardy, J. E. Littlewood, von Mangoldt, Charles de la Vallee Poussin, Carl Siegel, George Polya, Jacques Hadamard, Edmund Landau, Bohr, Atle Selberg, Emil Artin Andre Weil, Pierre Deligne, Jorgen Pedersen Gram, Alaine Connes, Ernst

124-Riemann Hypothesis and PNT

Lindelof, Harald Cramer. It influenced other computing professionals such as Alan Turing, Andrew Odlyzko, Freeman Dyson and Hugh Montgomery. Though the roots of conjecture can be well traced to Euler, Legendre and Gauss as they paid quiet a bit attention to the theory behind the conjecture. The great source of importance of Riemann conjecture lies in understanding the properties surrounding the prime numbers and hence the computational algorithms of major importance.

Getting to the formula relating zeta function to prime numbers:

$$\zeta(s) = \sum_{n=1}^{\infty} \frac{1}{n^s} = \prod_{p} \frac{1}{\left(1 - \frac{1}{p^s}\right)} \tag{1}$$

Where n ranges over all positive integers (n=1,2,3,......) and p ranges for all prime numbers (p=2,3,5,7,11,........).Equation (1) is the core function for showing the relationship between the zeta function and prime numbers. For s as even 2,4,6,.........∞, the closed form of series known to be first found by Euler in 1734. However, for the odd powers, the closing value remained conjecture till the date a proof has been presented in this book. Riemann extended the value of s

App. 9 Riemann Zeta Function- Analysis 2-125

to complex number and to the global variable through the functional equation. Riemann functional equation did not envisage any importance to Apery constant and odd zeta function in general.

Riemann and mathematicians whose work were inspired by Riemann assumed the very first principles of finding the value of zeta function when s is odd, which could have shed light on complex numbers used in Riemann Zeta function or anywhere in the analytical number theory. The inquiry of Gauss for quadratic reciprocity theorem laid foundation for finding the closed value of Apery constant and other odd powered zeta function.

Gauss established in 1849 that he believed as early as 1792 that the density of prime numbers is $1/\log(x)$ and published the paper in the following years confirms his belief in the accuracy of this approximation. The first significant result beyond Euler was obtained by Chebyshev in around 1950. Chebyshev proved that the relative error in the approximate integral.

126-Riemann Hypothesis and PNT

$$\frac{\pi(x)}{\int_{\infty}^{\infty} \frac{1}{\ln x} dx}$$

Chebyshev also showed that as the limit of x→∞, the principal result of the ratio shall tend to 1.

$$Li(x) = \int_2^x \frac{dt}{\log t} + Const$$

The error correction and corrected prime counting is presented in Chapter 5 of this book.

Riemann's function deals with $\zeta(s)$ as complex variable where s is the complex number, for the study of complex zeros of $\zeta(s)$.

Considering the right hand side of equation (1) we obtain

$$\frac{1}{\left(1-\frac{1}{p^s}\right)} = 1 + \frac{1}{p^s} + \frac{1}{(p^2)^s} + \frac{1}{(p^3)^s} + \ldots$$

And thus its product is therefore the sum of the terms of form:

$$\frac{1}{\left(p_1^{n1} p_2^{n2} \ldots\ldots p_r^{nr}\right)}$$

By the fundamental theorem of arithmetic, every integer can be expressed as the product of prime in just exactly one way for sum is just $\sum(1/n^s)$. Euler used this formula primarily for the even values. At that time there was no idea for the closing values of $\sum(1/n^s)$ for odd s = 3,5,7,9...

Dirichlet gave rigorous proofs that the equation (1) can be extended for all real values of s. Dirichlet appears to have given no serious consideration to the idea of quadratic reciprocity and odd zetas functions. Riemann extended the values of s in equation (1) to complex number and he goes further to make the series diverge and to logically deduce the results from divergence. Riemann deduced that all the non trivial zeros of zeta function has real part ½ but did not prove it. Since then many mathematicians have tried to prove the theorem that all the non trivial roots of zeta function has real part ½ but without any success. Based on the real analysis it can be easily shown, as presented in Chapter 2 and Chapter 3 as why the real part of Riemann Zeta function is ½.

128-Riemann Hypothesis and PNT

Euler extended the idea of n! =n(n-1)(n–2)............3.2.1 from natural number to all real numbers or rather every complex numbers in half plane s>-1 :

$$n! = \int_0^\infty e^{-x} x^n dx$$

Gauss introduced the term $\prod(s)$ such that:

$$\prod(s) = \int_0^\infty e^{-x} x^s dx$$

$\prod(s) = s!$

when s is a natural number

$\prod(s)$ has no zeros is analytical function of complex variable s and has simple poles at s=-1,-2,-3...

However, it was later modified by Legendre and he introduced the symbol $\Gamma(s)$ in place of $\prod(s-1)$.

Therefore the modified gamma function is represented as:

App. 9 Riemann Zeta Function

$$\Gamma(z) = \int_0^\infty t^{z-1} e^{-t} dt \qquad Rz>0$$

Another definition of gamma function is given by Euler viz.

$$\Gamma(z) = \underset{n \to \infty}{Lim} \frac{n! \, n^z}{z(z+1)(z+2)\ldots\ldots(z+n)} \qquad (2)$$

$z \neq 0, -1, -2, \ldots\ldots$

Obviously, $\prod(z) = z! = \Gamma(z+1)$ for z is natural number and the following two trigonometric relationship of the gamma function viz.

$$\Gamma(\tfrac{1}{2}+x)\Gamma(\tfrac{1}{2}-x) = \frac{\pi}{\cos(\pi x)} \qquad (3)$$

$$\Gamma(x)\Gamma(1-x) = \frac{\pi}{\sin(\pi x)} \qquad (4)$$

Here, our point is as why the gamma function is unilateral (positive) boundary condition rather than universal (positive as well negative) boundary condition, is easily deduced from equation (2). These boundary conditions are well mentioned in the Chapter 3 "Closing value for Apery Constant".

130-Riemann Hypothesis and PNT

In Riemann Zeta Function, we are dealing with the laws of nature, as we previously dealt in the product formulation of the sine function of Euler's famous result.

Why the laws of nature (related to numbers) are partially baffled by the gamma function and all the subsequent applications in the results of Riemann Hypothesis and Prime Number theorem? Let's compare this with the closing value of Apery Constant in the Chapter 3. In gamma function we are just dealing with half the boundary conditions and totally omitting the full scale consideration required for bringing gamma function in the derivation of functional equation. Are the infinite number of trivial zeros of functional equation derived by Riemann is taken for granted presence and assume cause no error?

Another way to say is that the simple poles s = -1, -2, -3,.......... ∞ terms have effect on the exactness of the solution. The polynomial way to express these poles is:

$$(1+x)\left(1+\frac{x}{2}\right)\left(1+\frac{x}{3}\right)\ldots\ldots\ldots\ldots$$

App. 9 Riemann Zeta Function- Analysis 2-131

The Riemann's basic philosophy was to deal with analytical function globally in terms of power series. Riemann derives his formula for $\zeta(s) = \sum n^{-s}$ for which shall "remain valid for all s"

Riemann's formula for $\zeta(s)$ takes the form:

$$\zeta(s) = \frac{\Pi(-s)}{2\pi i} \int_{\infty}^{\infty} \frac{(-x)^s}{e^x - 1} \frac{dx}{x} \qquad (5)$$

For real value of s>1,

$$\zeta(s) = \sum_{1}^{\infty} \frac{1}{n^s} \qquad (6)$$

As per equation (5), $\zeta(s)$ shall have the simple poles at s=1, 2, 3... which are complex analytical to $\Pi(-s)$ poles, but according to Remainder theorem or Euler's derivation of equation (6) for even powers, there are no poles at s=2, 3, 4,..... But of course there is a pole at s=1 for which $\zeta(s) = \infty$. Therefore, the equation (6) is analytical at every point except at s=1; for which there is a pole.

132-Riemann Hypothesis and PNT

Recalling the Bernoulli numbers B_k defined by:

$$\frac{z}{e^z-1} = \sum_{k=1}^{\infty} \frac{B_k z^k}{k!} \quad (z < 2\pi) \tag{7}$$

In fact the formula

$$\zeta(2k) = \frac{(-1)^{k+1}(2\pi)^{2k} B_{2k}}{2(2k)!} \tag{8}$$

can be derived by substituting $x = -\frac{1}{2}\,ir$ into the famous identity:

$$\sin x = x \prod_{k=1}^{\infty}\left(1 - \frac{x^2}{\pi^2 k^2}\right) \tag{9}$$

By logarithmic differentiation one obtains:

$$\frac{1}{e^r-1} = \frac{1}{r} - \frac{1}{2} + \sum_{k=1}^{\infty} \frac{2r}{4k^2\pi^2 + r^2}$$

So that $B_0 = 1$, $B_1 = -\frac{1}{2}$ and $B_{2k+1} = 0$ for $k \geq 0$
Or, in general

$B_0 = 1$ $B_1 = -1/2$

$B_2 = 1/6$ $B_3 = 0$

$B_4 = -1/30$ $B_5 = 0$

$B_6 = 1/42$ $B_7 = 0$

$B_8 = -1/30$ $B_9 = 0$

The odd Bernoulli numbers are all zero. For even Bernoulli i.e. B_{2k} there is no simple computational formula for them.

Riemann's zeta functional equation in terms of Gauss's gamma definition is:

$$\zeta(s) = \Pi(-s)(2\pi)^{s-1} 2 \sin(s\pi/2) \zeta(1-s) \tag{10}$$

Or,

$$\Pi\left(\frac{s}{2}-1\right)\pi^{-s/2}\zeta(s) = \Pi\left(\frac{1-s}{2}-1\right)\pi^{-(1-s)/2}\zeta(1-s) \tag{11}$$

This means the function on left side of (11) does not change by substituting s = 1-s.

When sin $\pi x = 0$, x has poles at $\pm 1, \pm 2, \pm 3$...

Similarly, when cos $\pi x = 0$, x has poles at $\pm ½, \pm 3/2, \pm 5/2$...

It may be noted that cos $\pi x = [\cos(\pi x/2) - \sin(\pi x/2)][\cos(\pi x/2) + \sin(\pi x/2)]$ and hence the resulting infinite series is of the form of inverse even power in terms 2n-1 lattice points.

Now we definitely know, $\zeta(2) = \pi^2/6$ $\zeta(4) = \pi^4/90$ and so on .There was conjecture that no one has so far succeeded in obtaining the formula as simple as equation (9) for $\zeta(2k+1)$. In fact besides the results that $\zeta(3)$ is irrational almost nothing was known about the arithmetical structure of $\zeta(2k+1)$ for k>1; till now since the proof is presented in this book.

If we use the modern notation for Gamma Function viz.; for every Re s> 1; the Gamma Function is defined by:

$$\Gamma(s) = \int_0^\infty e^{-x} x^{s-1} dx \qquad (12)$$

and for other values by analytical continuation. The Gamma Function is a regular function of s in whole plane s= 0, -1, -2, -3..., are the poles of first order with residues $(-1)^n/n!$

Where n = 0, 1, 2, 3...

Some common properties of Gamma Functions are:

$$\Gamma(s+1) = s\Gamma(s) \qquad (13)$$

$$\Gamma(s)\Gamma(1-s) = \pi/\sin(\pi s) \qquad (14)$$

$$\Gamma(s)\Gamma(s+\tfrac{1}{2}) = 2\sqrt{\pi} 2^{-2s} \Gamma(2s) \qquad (15)$$

$$B(a,b) = \int_0^1 x^{a-1}(1-x)^{b-1} dx = \frac{\Gamma(a)\Gamma(b)}{\Gamma(a+b)} \quad (For\, Rea > 0, Reb > 0) \qquad (16)$$

$$\Gamma'(1) = \int_0^\infty e^{-x} \log x\, dx = -\gamma = -0.5772157 \qquad (17)$$

The γ in equation (17) is known as Euler's constant. In fact there are many properties Euler's constant listed in the

Appendix13. Accordingly, the Riemann's functional equation (10) becomes:

$$\pi^{-s/2}\Gamma(s/2)\zeta(s) = \pi^{-(1-s)/2}\Gamma[(1-s)/2]\zeta(1-s) \qquad (18)$$

The functional equation (18) represents one of the fundamental results of zeta function theory. It was discovered by B. Riemann and characterizes the philosophy of Riemann for global variables consideration therefore represents the most remarkable achievements. The functional equation evidently shows that $\zeta(-2n) = 0$ for n to be positive integer i.e. n=1, 2, 3..., since the gamma function has poles at negative integers and the pole of $\zeta(1-s)$ cancels the poles of $\Gamma(s/2)$. The cancellation of these poles have a negative effect on true proof based on quadratic reciprocity which is the source of error in developing true form of Prime Number Theorem.

In fact dividing equation (18) by $\Gamma(s/2)$ and limiting s $\to 0$ yields; $\zeta(0) = -½$.

The zeros at s = -2n are called "trivial zero". For the formula for $\zeta(-2n)=0$ concludes upon the fact that the odd Bernoulli numbers B_3, B_5, B_7,..... are all 0.

In literature there exist many proofs of functional equation. The function $\zeta(s)$ has infinity of complex zeros whose precise location is conjectured at the critical line $\sigma = \frac{1}{2}$ and the distribution of the zero represents major problems in zeta function theory. But since the functional equation does not hold the accountability of its infinite number of trivial zeros, it has not resulted in exact Prime Number Theorem or even for the idea of non-trivial zeros relationship with PNT.

Using the standard properties of Gamma Function, there are many ways for the equivalence of the functional equation and can be represented as:

$$\zeta(s) = \chi(s)\zeta(1-s), \quad \chi(s) = \frac{(2\pi)^s}{2\,\Gamma(s)\cos(\pi s/2)} \tag{19}$$

This is equivalent to:

$$\zeta(s) = 2^s\,\pi^{s-1}\sin(\frac{\pi s}{2})\Gamma(1-s)\zeta(1-s) \tag{20}$$

The functional equation developed by Riemann does not cover the boundary conditions for finding the closing value

138-Riemann Hypothesis and PNT

of Apery constant or odd zetas. The reason for this is that the functional equation has trivial zeros at -2, -4, -6...and the polynomial of which is:

$$\left(1+\frac{x}{2}\right)\left(1+\frac{x}{4}\right)\left(1+\frac{x}{6}\right)\ldots\ldots$$

Referring to the Chapter on Closing Value of Apery Constant, we note that the Riemann's functional equation just attempts to satisfy half way of the boundary value and hence is not valid for computation purposes.

The Hadamard Product Formula:

From the literature, it's worthwhile considering that the only zeros outside the critical strip $0 < \sigma < 1$ are so called trivial zeros s = -2n (n= 1,2,3,.....). the strip $1<\sigma<0$ is known as "critical strip" and the line $\sigma=½$ is known as critical line in the zeta function theory. The first product representation (22) of $\xi(s)$ was published by Hadamard in 1893. One of the major themes of Riemann's work is the global characterization of analytical functions by their singularities.

The Riemann - Von Mangoldt formula

The formula was conjectured by B. Riemann in 1859 and was proved by H. von Mangoldt around 1905. It furnishes precise expression for number of zeros N(T) of the Riemann Zeta function $\zeta(s)$ in the region $0<\sigma<1$, and $0<t\leq T$ and thus represents an important analysis tool in the Riemann Zeta function theory.

$$N(T) = \frac{T}{2\pi}\log\frac{T}{2\pi} - \frac{T}{2\pi} + O(\log T) \qquad (21)$$

In fact these numbers of roots can be explained in terms of re normalization theory. Riemann considers the left side of the functional equation, which occurs in the symmetrical form of the functional equation remains unchanged when s is replaced by 1-s has its poles at s =0 and s =1. Riemann multiplies this function by s(s-1)/2 and defines

$$\xi(s) = \Pi(s/2)(s-1)\pi^{-s/2}\zeta(s) \qquad (22)$$

$\xi(s)$ is considered to be analytical at every point over s. Riemann's goal was essentially to prove

$$\xi(s) = \xi(0)\prod_{\rho}\left(1-\frac{s}{\rho}\right) \tag{23}$$

Where ρ ranges over the roots for ξ(ρ) = 0

If one considers the analogy of the expression of sin (x) as an infinite degree polynomial expressed as Euler's famous equation and product formula (10); it can be easily perceived that ξ(s) is the a polynomial of infinite degrees. But again equation (23) misses the critical boundary conditions required for finding the closing values of Apery constant and odd zetas in general. It just attempt to satisfy half the boundary condition of the integrated whole for the reciprocity sign requirement, and as per Gauss for the deep properties of prime numbers and therefore of the zeta function, which requires the positive and negative infinite product for appropriate boundary conditions of the odd zetas.

The functional equation, the gamma function and even anything related to e=2.71...do not satisfy the comprehensive boundary condition required for finding the closing values of Apery constant. In fact, exponential function e= 2.71……, by definition just contains the positive infinite product and therefore misses out the boundary conditions for quadratic

App. 9 Riemann Zeta Function- Analysis 2-141

reciprocity and according to Gauss's theorem for error evaluation because of odd zetas.

According to Riemann's paper and subsequent records has unproved hypothesis that the equation $\xi(\frac{1}{2} + it) = 0$ has approximately $(T/2\pi)\log(T/2\pi)$ real roots α in the range of $0 < \alpha < T$

The original question for the validity of $\pi(x) \approx \int_{2}^{x} \frac{dt}{\log t}$ remains unresolved by Riemann's paper. However, in 1896 Hadamard and de la Vallée Poussin independently proved the error in the approximation does approach 0 as the limit x → ∞. But that is just an approximation.

The location of roots:
The theoretical calculations and location of the roots can be laid out as Euler Maclaurin Summation. The first substantial finding of roots was presented by Grams in 1903 and published 15 roots in Re s = ½. Typically the values of the roots he gave were:

$\rho = \frac{1}{2} + i\alpha$

142-Riemann Hypothesis and PNT

The number of first n zeros that are satisfy $t > 0$, $\sigma = \frac{1}{2}$ was historically established as follows:

J. Gram (1903)	N=15
R. Backlund (1914)	N=79
J.I. Hutchinson (1925)	N= 138
E.C. Titchmarch (1935)	N= 1,041
D.H. Lehmer (1956)	N= 25,000
N.A. Meller (1958)	N= 35,337
R. Sherman Lehman (1966)	N= 250,000
J.B. Rosser et. Al. (1969)	N= 3,500,000
R.P. Brent (1979)	N= 81,000,001
J. van de Lune et. al. (1981)	N= 200,000,001
J. van de Lune et. al. (1983)	N= 300,000,001

Supposing we need to calculate $\zeta\left(\frac{1}{2} + 14.0i\right)$ which is:

$$\zeta(\tfrac{1}{2} + 18i) = \frac{1}{1^{(\frac{1}{2}+18i)}} + \frac{1}{2^{(\frac{1}{2}+18i)}} + \frac{1}{3^{(\frac{1}{2}+18i)}} + \ldots\ldots\ldots \infty$$

By calculation from Euler Maclaurin Summation formula we obtain;

$\zeta(½ + 14.0i) = +.02 - 0.10i$

By an increment we obtain:

$\zeta(½ + 14.2i) = -0.01 + 0.05i$

Therefore one observes the sign changes from positive i.e. $\zeta(½+ 14.0i)$ to negative i.e. $\zeta(½ + 14.2i)$ giving us the indication that there is 0 of zeta function along the critical line $\sigma = ½$ which comes out to be $\alpha = 14134\ 725$

We will show at later chapters as how Euler Maclaurin misses out the boundary condition for the closing value of Apery constant. This is because Euler's Maclaurin summation uses only Bernoulli's number and not Euler's number or the Euler's extension number.

The error in complex analysis of Riemann Zeta function can be explained in terms cube roots of unity, which are 1, and two conjugate complex roots i.e. $\dfrac{-1 \pm \sqrt{3}i}{2}$.

144-Riemann Hypothesis and PNT

Appendix 10
Riemann Zeta Function Analysis Elements of Computations

The 88% rise in Microsoft stock in 1996 meant [Bill Gates] made on paper more than 10.9 billion. That makes him world richest person, by far. But he is more than that, a technologist turned entrepreneur, he embraces the digital era.

- Walter Isaacman (1952-)
"In search of real Bill Gates"
Time January 1992

We now move to the computation aspect of Riemann Hypothesis. As determined by us the formula for the number of prime number less than a given number is Li(x). Depends on which factor is greater than which the Gaussian adjustments shall be either 3 mod 4 or 1 mod 4. The formula makes correction over Li(x) is expressed as:

$$Li(x) = \int_{2}^{\infty} \frac{1}{\log x} dx$$

Subsequent corrections were made by S. Ramanujan.

Our modified formula for the prime number less than a given number is given and expressed as Modified Li(x) or simply Mli(x):

$$MLi(x) = \int_{2}^{\infty} \frac{1}{\log(x - \epsilon)}$$

Where ϵ is the modified factor

The modified factor ϵ obviously depends on errors caused by odd zeta function. As we have noted from the previous chapters that the odd zeta functions i.e. $\zeta(3)$, $\zeta(5)$, $\zeta(7)$,... are not purely cyclic ; i.e. they cannot be purely expressed as the rational multiple of the powers of π as in even zeta function $\zeta(2)$, $\zeta(4)$, $\zeta(6)$...which can be expressed as the rational multiple of even powers of π. This also brings the issue of Chinese Remainder, Wilson Theorem, Euler's Totient function and Fermat's Little theorem.

The computation for determining the number of prime numbers less that a given number may be tedious. The reason for this being that computers are not very exact with floating point numbers and the precision is limited by computer architecture. Computers can compute and represent the integers with precision but the overflow occurs for reasonably large floating point computations although with the augment of virtual memory the integer computation can be accurately represented.

For example, the internal memory of 32 bit computer can hold maximum signed integer 2147483647 and can hold minimum integer -2147483648 and can hold maximum unsigned integer 4294967295. In this case any computation dealing with integers greater than these or less than the minimum integer shall produce error message unless we use the virtual memory and routines such C++ big integer or Java Big Integer or PERL Big Integer. Information about Java Big Integer is available at the Sun-microsystem web

148-Riemann Hypothesis and PNT

page, for PERL at CPAN. In 32 bit windows programming Large Integer is sometimes used by the experienced windows programmer.

There are many C++ compilers for the computations are available at the web. Some are expensive, some are less expensive and some are totally free. If you buy, just buy the least expensive ones available over the web. Windows does not come with the built in C++ compiler. But the public Universities and other industrial organizations may have compiler installed to it. Some examples of the compilers are Microsoft Visual Basic C++, Code Warriors or Borland Compiler. Some of these compilers have their own text editor and debugger. In other the programs can just be typed in the notepad or the WordPad editor which usually comes with windows operating systems.

In Unix workstation the compiler may have already been installed and the program with the filename.c shall compile with the simple command cc. filename.

The GNU foundation has free C++ compiler available along with their emacs editor and build in software for Unix and DOS can be downloaded from their web site.

There are many good books available for numeric computations. It's a good idea to start with GNU scientific library. Another language of interest is Perl as it is beneficially used in computational intensive field of bio-informatics.

As human we are used to think in terms of decimal (base 10). But computers are designed on binary (0 or 1). It was easy because any switching on the part of transistor represent only two states off (0) or on (1).

The computer architecture mainly consists of memory arranged sequentially i.e. 1,2,3,4 ... A bit is the simple state of 0 or 1. A byte is composed of 8 bits. Depending on number of bytes/ bits the computer operating systems is 32 bits (4 bytes) or even 64 bits (8 bytes). Once we have the information expressed in binary, it is pretty easy to convert into another form e.g. hexadecimal, octal etc.

Computer essentially consists of Central Processing Unit, Main Memory, Input Device, Output Device, Hard Drive. A computer can process information up to several MHz. If the four bytes are assigned to single 32 bit integer in C++, the single decimal integer represented shall be 1,115,251, 968 as

150-Riemann Hypothesis and PNT

$01000010011110010110010100000000_2 = 1{,}115{,}251{,}968_{10}$

If the same four bytes are assigned to floating point the algorithm is bit complicated.

It divides the 32 bits string in three parts:
(1) Left most bit is the sign bit
(2) The next 8 bits are exponent
(3) The right-most 23 bit are fraction.

The sign bit is 0. The value of exponent is $10000100_2 - 127 = 132 - 127 = 5$

The fraction value is $1.11110010110010 1_2$

Therefore the complete floating point number is :

$1.11110010110010 1_2 \times 2^5 = 111110.00101100101_2 = 2^5 + 2^4 + 2^3 + 2^2 + 2^1 + 2^{-2} + 2^{-4} + 2^{-5} + 2^{-8} + 2^{-10} = 62.34863$

Note that 127 is subtracted from 8-bit exponent and 1 is added to 23-bit fraction. This algorithm is called excess 127 floating point representation.

APPENDIX 11
Ramanujan Odd Zeta Function Analysis

The Riemann Zeta function is given by well known identity i.e.:

$$\zeta(s) = \prod_{n=1}^{\infty} \left(1 - p_n^{-s}\right)^{-1} \qquad (1)$$

The equation (1) can also be written as:

$$\zeta(s) = \left(1 - 2^{-s}\right)^{-1} \prod_{m \equiv 1 (\bmod 4)} \left(1 - m^{-s}\right)^{-1} \prod_{n \equiv 3 (\bmod 4)} \left(1 - n^{-s}\right)^{-1} \qquad (2)$$

Where m and n are primes congruent to 1 and 3 modulo 4, respectively

For even $n \geq 2$ \qquad (3)

$$\zeta(n) = \frac{2^{n-1}|B_n|\pi^n}{n!}$$

Where B_n are Bernoulli numbers

Another relationship between the zeta function and Bernoulli's numbers is given by:

$$B_n = (-1)^{n+1} n\zeta(1-n) \qquad (4)$$

As $B_n = 0$ trivially when n is odd, the equation (4) becomes:

$$B_n = -n\zeta(1-n) \qquad (5)$$

Rewriting (5) yields,

$$\zeta(-n) = B_{n+1}/(n+1) \qquad (6)$$

Where B_n is Bernoulli's number and for few values of n=1, 3, 5... the Bernoulli's number B_n is given as -1/12, 1/120, -1/252... And in general as shown in Appendix From the deep results of the renormalization theory (Elizade 1995, Block 1996, Lepowski 1999), using Kronecker Delta one can obtain the results $\zeta(0) = -1/2$; $\zeta(-1) = -1/12$

Analytical form for $\zeta(2n+1)$ is not known till now but it is somewhat conjectured that $\zeta(3)$ can be written as:

$$\zeta(3) = \frac{1}{2}\sum_{k=1}^{\infty}\frac{1}{k^2}\left(1+\frac{1}{2}+\ldots+\frac{1}{k}\right) = \frac{1}{2}\sum_{k=1}^{\infty}\frac{H_k}{k^2} \qquad (7)$$

Appendix 11 Ramanujan Zeta Function-153

$\zeta(n)$ can also be expressed as:

$$\zeta(n) = x \xrightarrow{Lim}_{\infty} \frac{1}{(2x+1)^n} \sum_{k=1}^{x} \left[\cot\left(\frac{k}{2x+1}\right)\right]^n \qquad (8)$$

for n=3, 5, 7,……..

For Mobius function μ(n), the relation with zeta function is:

$$\frac{1}{\zeta(s)} = \sum_{n=1}^{\infty} \frac{\mu(n)}{n^s}$$

Srinivas Iyengar Ramanujan was the first person to discover the rapidly converging series for $\zeta(n)$ for the odd n.
For n>1 and n≡3(mod4)

$$\zeta(n) = \frac{2^{n-1} \pi^n}{(n+1)!} \sum_{k=0}^{(n+1)/2} (-1)^{k-1} \binom{n+1}{2k} B_{n+1-2k} B_{2k} - 2\sum_{k=1}^{\infty} \frac{1}{k^n (e^{2\pi k} - 1)} \qquad (9)$$

where B_k is Bernoulli's number.
For n≥5 and n≡3(mod4) the corresponding formula Ramanujan gave:

154-Riemann Hypothesis and PNT

$$\zeta(n) = \frac{(2\pi)^n}{(n+1)(n-1)} \sum_{k=0}^{(n+1)/4} (-1)^k (n+1-4k)\binom{n+1}{2k} B_{n+1-2k} B_{2k} \quad (10)$$

$$-2\sum_{k=1}^{\infty} \frac{e^{2\pi k\left(1+\frac{4\pi}{k-1}\right)-1}}{k^n \left(e^{2\pi k}-1\right)^2}$$

By defining

$$S\pm(n) = \sum_{k=1}^{\infty} \frac{1}{k^n \left(e^{2\pi k} \pm 1\right)}$$

one obtains:

$$\zeta(3) = \frac{7\pi^3}{180} - 2S_-3 \quad (11)$$

$$\zeta(5) = \frac{1}{294}\pi^5 - \frac{72}{35}S_-(5) + \frac{2}{35}S_+(5) \quad (12)$$

$$\zeta(7) = \frac{19}{56700}\pi^7 - 2S_-(7) \quad (13)$$

$$\zeta(9) = \frac{125}{3704778}\pi^9 - \frac{992}{495}S_-(9) - \frac{2}{495}S_+(9) \quad (14)$$

$$\zeta(11) = \frac{1453}{425675250}\pi^{11} - 2S_-(11) \quad (15)$$

Appendix 11 Ramanujan Zeta Function-155

$$\zeta(13) = \frac{89}{257432175}\pi^{13} - \frac{16512}{8255}S_-(13) - \frac{2}{8255}S_-(13) \qquad (16)$$

$$\zeta(15) = \frac{13687}{390769879500}\pi^{15} - 2S_-(15) \qquad (17)$$

$$\zeta(17) = \frac{397549}{112024529867250}\pi^{17} - \frac{261632}{130815}S_-(17) - \frac{2}{130815}S_+(17) \qquad (19)$$

$$\zeta(19) = \frac{7708537}{21438612514068750}\pi^{19} - 2S_-(19) \qquad (20)$$

$$\zeta(21) = \frac{68529640373}{1881063815762259253125}\pi^{21} - \frac{4196352}{2098175}S_-(21) - \frac{2}{2098175}S_+(21) \qquad (21)$$

Since the derivation of Ramanujan odd powered is based on e= 2.71.....which is true for global consideration, but is not applicable for the global treatment of discreet logarithmic function because of the missing of global boundary conditions required for the Euler's product formula. The true value of discreet logarithmic function is dependent on the boundary conditions of positive and negative infinite products and according to Gauss theorem for the sign factor of the the reciprocity of prime numbers. Since these derivations are based on e = 2.71..... which is just a positive infinite product, merely half of the boundary condition of integrated whole is considered causing the results to be based on certain modulus form only and the Gauss's reciprocity theorem and thus the deep properties of prime number are missed in the Ramanujan odd zeta functions.

The eternal mystery of the world is its comprehensibility.

- Albert Einstein (1879-1955)
Journal of the Franklin Institute
March 1936

Appendix 12

An Elliptic Form of Infinite Series

As we know that Taylor Series Expansion of Sec (x) and Sech (x) leads to our understanding of Euler Numbers while Bernoulli's Number are self evident expansion of exponential function.

However there exists certain elliptic function such as Catalan's Constant which is defined as:

$$\frac{1}{2}\int_0^1 x\,dx = \frac{1}{2}\int_{x=0}^1 \int_{\theta=0}^{\frac{\pi}{2}} \frac{d\theta\,dx}{\sqrt{1-x^2 \sin^2\theta}} = \frac{1}{1^2} - \frac{1}{3^2} + \frac{1}{5^2} - \ldots\ldots\infty = 0.9159655\ldots\ldots \quad (1)$$

The series (1) is unique in the sense that it cannot be derived within the entire gamut of Cartesian coordinate system unless employed derivation techniques of elliptical function. The series such as

158-Riemann Hypothesis and PNT

$$1+\frac{1}{3^2}+\frac{1}{5^2}+\frac{1}{7^2}+\ldots\ldots\infty = \frac{\pi^2}{12}$$

can be generated using $\cos(x)=0$

Series such as

$$1-\frac{1}{3}+\frac{1}{5}-\frac{1}{7}+\ldots\ldots\infty = \frac{\pi}{4}$$

can be generated by equation $\cos(x/2)-\sin(x/2)=0$.

Noting that $\cos(x)=[\cos(x/2)-\sin(x/2)][\cos(x/2)+\sin(x/2)]$; but this equation does not derive the Catalan's constant.

The algebraic derivation of Catalan's constant even if it incorporates the real and imaginary roots, i.e. the root generating function generates the roots of the form which is beyond the scope of any root generating functions in the Cartesian coordinate system!

The series generated by algebraic equation $\sin(x) = 0$ is as

$$1+\frac{1}{2^2}+\frac{1}{3^2}+\frac{1}{4^2}+\ldots\ldots\infty = \frac{\pi^2}{6}$$

The $\sin(x)=0$ generate the above series or any even powered series is just equivalent to the generation of series by equation $\cos(x) = 0$

$$\sum_{k=1}^{\infty} \frac{1}{k^n} = \left[1 - \frac{1}{2^n}\right] \sum_{k=1}^{\infty} \frac{1}{k^{2n-1}}$$

As sin(x) and cos(x) are orthogonal functions, which means the cycle is complete within orthogonal function. But things are different for odd powered series and in this case the $\sin(x) = 0$ and $\cos(x) = 0$ do not behave as orthogonal generating functions.

$$1 + \frac{1}{3^3} - \frac{1}{5^3} - \frac{1}{7^3} + + - - \ldots \infty = \frac{3\pi^3 \sqrt{2}}{128} \tag{2}$$

which is the simple extension of:

$$1 - \frac{1}{3^3} + \frac{1}{5^3} - \frac{1}{7^3} + \ldots \infty = \frac{\pi^3}{32} \tag{3}$$

Or, the most general case

$$\frac{1}{1^{2p+1}} - \frac{1}{3^{2p+1}} + \frac{1}{5^{2p+1}} - \frac{1}{7^{2p+1}} + \ldots \infty = \frac{\pi^{2p+1} E_p}{2^{2p+2} (2p)!}$$

160-Riemann Hypothesis and PNT

where E_p is an Euler Number.

The sums of powers of positive integers are represented by:

$$1^p + 2^p + 3^p + \ldots + n^p = \frac{n^{p+1}}{p+1} + \frac{1}{2}n^p + \frac{B_1 p n^{p-1}}{2!} - \frac{B_2 p(p-1)(p-2)n^{p-3}}{4!} + \ldots$$

Where B_k are the Bernoulli numbers and this series terminates at n^2 or n according to p is odd or even.

Some of the special cases are:

$$1 + 2 + 3 + 4 + \ldots + n = \frac{n(n+1)}{2}$$

$$1^2 + 2^2 + 3^2 + 4^2 + \ldots + n^2 = \frac{n(n+1)(n+2)}{6}$$

$$1^3 + 2^3 + 3^3 + 4^3 + \ldots + n^3 = \frac{n^2(n+1)^2}{4} = (1 + 2 + 3 + 4 + \ldots + n)^2$$

$$1^4 + 2^4 + 3^4 + 4^4 + \ldots + n^4 = \frac{n(n+1)(2n+1)(3n^2 + 3n - 1)}{30}$$

App. 12 Elliptic Infinite Series-161

For general case;

$$S_k = 1^k + 2^k + 3^k + 4^k + \ldots + n^k$$

$$\binom{k+1}{1}S_1 + \binom{k+1}{2}S_2 + \ldots + \binom{k+1}{k}S_k = (n+1)^{k+1} - (n+1)$$

n and k are positive integers,

Adding equations (2) and (3) yields:

$$1 - \frac{1}{7^3} + \frac{1}{9^3} - \frac{1}{13^3} + \frac{1}{15^3} - \ldots \infty = \frac{(4 + 3\sqrt{2})\pi^3}{256}$$

In fact for any series of the form:

$$1 + \sum_{n=2, p=1}^{\infty} p, n \in \mathrm{N} \left[(-1)^{2p+1} \frac{1}{(2n-1)^{2p+1}} + \frac{1}{(2n+1)^{2p+1}} \ldots \right]$$

has the closed form in Cartesian coordinate system by the virtue of cyclic generating functions such as [a.cos(x) + b.sin(x)] or [a.cos(x) - b.sin(x)] and their combinations.

We observe that there are infinite Euler extension numbers and there are infinite odd zeta functions with alternate positive and negative signs. The equation of type (1) falls outside the realm of Cartesian coordinate in system of integers or rational numbers. Therefore, its closing value can only be defined in terms of its converging infinite series and not π (e.g. the closing value of Apery constant in terms of other alternate signs infinite series)

APPENDIX 13

Some Formulas Connecting γ

There are multitude of integrals in which utilizes γ, some of those are listed below:

$$\int_0^\infty e^{-x} \ln x\, dx = -\gamma \tag{1}$$

$$\int_0^\infty e^{-x^2} \ln x\, dx = -\frac{\sqrt{\pi}}{4}(\gamma + 2\ln 2) \tag{2}$$

Soldner's result:

$$L(x) = \int_2^x \frac{1}{\ln x}\, dx$$

$$L(x) = \gamma + \ln \ln x + \sum_{r=1}^\infty \frac{\ln^r x}{rr!} \tag{3}$$

164-Riemann Hypothesis and PNT

Johann Soldner (1766-1833) gave the latest corrected value of γ and also the series expansion of Li(x).

γ = 0.577 215 664 901 532 860 6065.......

$$\gamma = \int_0^1 \frac{1-e^{-u}-e^{-\frac{1}{u}}}{u} du \tag{4}$$

$$\int_0^1 \ln\ln\frac{1}{x} dx = -\gamma \tag{5}$$

$$\int_0^1 \frac{1}{\ln x} + \frac{1}{1-x} dx = \gamma \tag{6}$$

$$\int_0^\infty e^{-x} \ln^2 x \, dx = \frac{\pi^2}{6} + \gamma^2 \tag{7}$$

$$\sum_{r=2}^\infty \frac{\Lambda(r)-1}{r} = -2\gamma \tag{8}$$

where Λ is Von Mangold Function is defined as

App. 13- Some Formula connecting γ-

$$\Lambda(r) = \begin{cases} \ln p, & \text{if } r = p^m, p \text{ prime} \\ 0, & \text{otherwise} \end{cases}$$

$$n \xrightarrow{\lim} \infty \; \frac{1}{\ln n} \prod_{p \leq n} \left(1 - \frac{1}{p}\right)^{-1} = e^{\gamma} \tag{9}$$

$$n \xrightarrow{Lim} \infty \; \frac{1}{\ln n} \prod_{p \leq n} \left(1 + \frac{1}{p}\right) = \frac{6 e^{\gamma}}{\pi^2} \tag{10}$$

$$\int_0^\infty e^{-x} \left(\frac{1}{1 - e^{-x}} - \frac{1}{x}\right) dx = \gamma \tag{11}$$

$$n \xrightarrow{Lim} \infty \left(n - \Gamma\left(\frac{1}{n}\right)\right) = \gamma \tag{12}$$

$$\sum_{i=2}^{\infty} \frac{1}{i}(\zeta(i) - 1) = 1 - \gamma \tag{13}$$

166-Riemann Hypothesis and PNT

APPENDIX 14
Analysis of Riemann Zeta with Gamma Function

Man will occasionally stumble over the truth, but most time he will pick-up and carry on.

- Sir Winston Churchill

Riemann Zeta Function is defined by:

$$\zeta(x) = \frac{1}{1^x} + \frac{1}{2^x} + \frac{1}{3^x} + \frac{1}{4^x} + \ldots\ldots,$$

where x can be real integers greater than 1 or according to Riemann may be complex numbers of the form a + b*i*. In a very particular case, Riemann pointed in his paper that all the non trivial zeros (i.e. the transition from +ve to -ve value and vice versa) of the Riemann Zeta function lie on the straight line . It is defined in the other way viz.,

168-Riemann Hypothesis and PNT

$$\zeta(x) = \frac{1}{\Gamma(x)} \int_0^\infty \frac{u^{x-1}}{e^u - 1} du \quad x > 1$$

$$\zeta(1-x) = 2^{1-x} \pi^{-x} \Gamma(x) \cos(\pi/2) \zeta(x) \tag{1}$$

$$\zeta(2k) = \frac{2^{2k-1} \pi^{2k} B_k}{(2k)!} \quad k = 1, 2, 3, \ldots$$

The Gamma Function is defined as:

$$\Gamma(n) = \int_0^\infty t^{n-1} e^{-t} dt \quad for\ n > 0$$

Derivative of Gamma Function is defined as:

$$\Gamma'(1) = \int_0^\infty e^{-x} \ln x\, dx = -\gamma$$

$$\gamma = \lim_{n \to \infty} \left(1 + \frac{1}{2} + \frac{1}{3} + \frac{1}{4} + \ldots + \frac{1}{n} - \ln n\right) = 0.57721\ldots$$

γ is called Euler's constant

For n=0,1,2,3,......i.e. positive integers we have

$$\Gamma(n+1) = n!$$

and

App. 15 Analysis of Zeta with Gamma Func.-169

$\Gamma(n+1) = n\Gamma(n)$

The Gamma Function for n<0 can be defined by the equation:

$\Gamma(n) = \dfrac{\Gamma(n+1)}{n}$

The relationship between Gamma Function:

$\Gamma(x)\Gamma\left(x+\dfrac{1}{m}\right)\Gamma\left(x+\dfrac{2}{m}\right)\ldots\Gamma\left(x+\dfrac{m-1}{m}\right) = m^{1/2-mx}(2\pi)^{(m-1)/2}\Gamma(mx)$

For a special case m=2, the above equation reduces to:

$2^{2x-1}\Gamma(x)\Gamma\left(x+\tfrac{1}{2}\right) = \sqrt{\pi}\,\Gamma(2x)$

This is known as duplication formula

$\Gamma\left(m+\tfrac{1}{2}\right) = \dfrac{1.3.5\ldots(2m-1)\sqrt{\pi}}{2^m}$ for $m = 1, 2, 3, \ldots$

$\Gamma\left(-m+\tfrac{1}{2}\right) = \dfrac{(-1)^m 2^m \sqrt{\pi}}{1.3.5\ldots(2m-1)}$ for $m = 1, 2, 3, \ldots$

Other definitions and relationships of Gamma Function

170-Riemann Hypothesis and PNT

$$\Gamma(\tfrac{1}{2}) = \sqrt{\pi}$$

$$\Gamma(p)\Gamma(1-p) = \frac{\pi}{\sin p\pi}$$

Also,

$$|\Gamma(ix)|^2 = \frac{\pi}{x \sinh \pi x}$$

$$\Gamma(x+1) = \lim_{k \to \infty} \frac{1.2.3.4.......k}{(x+1)(x+2)(x+3).....(x+k)} k^x$$

$$\frac{1}{\Gamma(x)} = xe^{\gamma x} \prod_{m=1}^{\infty}\left[\left(1+\frac{x}{m}\right)e^{-x/m}\right]$$

This is an infinite product representation of Gamma function.

The Stirling Asymptotic Series for the asymptotic expansion of Gamma Function is given by:

$$\Gamma(x+1) = \sqrt{2\pi} x^x e^{-x}\left[1 + \frac{1}{12x} + \frac{1}{288x^2} - \frac{139}{51840x^3} + ...\right]$$

App. 15 Analysis of Zeta with Gamma Func.-171

For x = n, a positive integer, the useful expansion of n! for large n is given by Stirling Formula

$$n! \approx \sqrt{2\pi n}\, n^n e^{-n}$$

As is obvious from the definitions and formula of Riemann Zeta Function and the functional equation that the unilateral nature of gamma function and the omission of the quadratic residue because of the sign factor of reciprocity misses the deep properties governing the distribution of prime numbers. The trivial zero of the functional equation are all negative is another way to say that the condition for the reciprocity of the prime numbers are not met and the present equations though forming the equivalence of half boundary conditions, even in full boundary conditions does not represent anything more than the Euler's sine product formula which served as the basis for his derivation of famous result

$$\sum_{k=1}^{\infty} \frac{1}{k^2} = \frac{\pi^2}{6}$$

and all such even powered series viz. $\zeta(2n)$ where $n \in \mathbb{N}$. For deducing the closing value of odd zetas viz., $\zeta(2n+1)$ where $n \in \mathbb{N}$, requires both boundary conditions to be satisfied as per Gauss's Lemma or the reciprocity theorem for the signs of

172-Riemann Hypothesis and PNT

least residue.

As Riemann Hypothesis is concerned with the laws of nature of numbers, i.e. the universal distribution of the prime numbers, we cannot apparently neglect the subtle and inner effect boundary conditions as governed by the Gauss's reciprocity theorem. We cannot simply cancel the poles of $\zeta(1-x)$ with the poles of $\Gamma(x)$ in equation (1) and neither can we equate the half boundary condition poles.

Half the truth is often great lie

                     - Benjamin Franklin (1706-1790)
                     Poor Richard Almanack
                     July 1758

APPENDIX 15

Glossary

(1) Positive Infinite Product:

Positive infinite product is defined as:

$$\prod_{\substack{n=1 \\ x>0}}^{\infty}\left(1+\frac{x}{n}\right)$$

Examples of positive infinite products are gamma function, exponential function etc.

(2) Negative Infinite Product:

Negative infinite product is defined as:

$$\prod_{\substack{n=1 \\ x>0}}^{\infty}\left(1-\frac{x}{n}\right)$$

Examples of negative infinite product are gamma function of

174-Riemann Hypothesis and PNT

(-x), e.g. $\Gamma(-x)$, or inverse of exponential function, e^{-1} or e^{-x}

(3) The positive as well as negative infinite product:
An example of positive as well as negative infinite product is the infinite product form for sin(x) or cosine(x), for example

$$\sin x = x \prod_{n=1}^{\infty}\left[1 - \frac{x^2}{(n\pi)^2}\right]$$

It is worth noting that the above product formula for the sine function is derived from the premise that $\sin(x) = 0$ when $x \neq 0$ and therefore we can omit the 'x' before the product sign on the right hand side of the equation.

Because of the product form of the function $\sine(x) = 0$, we are in the position to infer the closing value of Apery constant and odd zeta functions in particular.

(4) Even zetas :
Even zeta function is defined as $\zeta(2n)$ where $n \in \mathbb{N}$
Examples are $\zeta(2)$, $\zeta(4)$, $\zeta(6)$,..........

(5) Odd zetas:

Odd zetas is defined as $\zeta(2n+1)$ where $n \in \mathbb{N}$

Examples are $\zeta(3), \zeta(5), \zeta(7),\ldots\ldots$

(6) Odd powered odd numbered alternate series:

The infinite series consisting of odd numbers type where it changes sign e.g.

$$1 - \frac{1}{3^3} + \frac{1}{5^3} - \ldots\ldots$$

(7) Boundary Conditions:

Boundary condition in zeta function refers to the extreme boundary conditions in the event of random products of positive and negative infinite products. For details, refer to the Chapter 3. -The Closing Value of Apery constant.

(8) Euler's Extension number:

Bernoulli's number and Euler's number are very common in mathematics literature. By little discipline, it may be observed Bernoulli's number are associated with even zetas, i.e. $\zeta(2), \zeta(4), \zeta(6),........$ and interestingly Euler numbers are **not** associated with odd zetas or $\zeta(3), \zeta(5), \zeta(7),.....$; but are associated with only particular type of infinite series viz.

$$1 - \frac{1}{3^{(2n+1)}} + \frac{1}{5^{(2n+1)}} - \frac{1}{7^{(2n+1)}} + ...\infty \; terms \tag{1}$$

where $n \in \mathbb{N}$

Series (1) can be easily derived from infinite product using the remainder or factorial theorem on equation:

$\cos(x) - \sin(x) = 0$

or

$\cos(x) + \sin(x) = 0$

In fact there are infinite number of series with odd power and alternate signs of (+)ve and (-)ve can be derived because $a.\cos(x) + b.\sin(x) = 0$ can be expressed as $\cos(x-\alpha)$ where

$$\alpha = \sin^{-1}\left(\frac{a}{\sqrt{a^2+b^2}}\right) = \cos^{-1}\left(\frac{b}{\sqrt{a^2+b^2}}\right)$$

Therefore, the numbers associated with other such type of alternate sign infinite series is named as Euler's extension number. Example of such infinite number of such series is:

$$1 - \frac{1}{7^{(2n+1)}} + \frac{1}{9^{(2n+1)}} - \frac{1}{15^{(2n+1)}} + \frac{1}{17^{(2n+1)}} - \ldots \ldots \infty \text{ terms}$$

where $n \in \mathbb{N}$

(9) Residue of the function:

Every mathematical function has residue.

Mathematically,

$\text{Res } F(x)_{x=a} \equiv$ Residue of function $F(x)$ at point 'a'

178-Riemann Hypothesis and PNT

A function F(x) is said to be continuous at point 'a', if :

$$\frac{d(F(x))_{x=a}}{dx} \neq 0$$

otherwise discontinuous.

Since calculus is based on positive feedback, i.e.

$$\frac{d(F(x))}{dx} = \Delta x \xrightarrow{Lim} 0 \left[\frac{(x+\Delta x) - x}{\Delta} \right]$$

We miss the negative feedback, i.e.

$$\frac{d(F(x))}{dx} = \Delta x \xrightarrow{Lim} 0 \left[\frac{(x-\Delta x) - x}{\Delta x} \right]$$

For the numeric lattice point, we need to consider both positive and negative feedback with appropriate boundary conditions as defined by the Closing Value of Apery constant (Chapter 3) or Gauss Theorem (Appendix 2)

(10) Factorial or Remainder Theorem:

This is the very basic and important theorem.

For a polynomial of degree n, viz.:

$F(x) = a_0 + a_1 x + a_2 x^2 + \ldots a_n x^n$
where $a_1, a_2, \ldots a_n$ are coefficients

If F(a) = 0 then (x-a) is the root of F(x)

Or, (x-a) is one of the factors of F(x)

(11) Euler's Totient Function Φ :

The Euler's Totient function is defined as-

Given a number 'n', the Φ are the numbers less than 'n' which are relative prime to 'n'.

e.g. consider a number 35 which can be represented as the products of two prime numbers 7 x 5. The numbers less than 35 which are relative prime to 35 is (7-1).(5-1) = 6x4 = 24.

On the basis of fundamental theorem of arithmetic we can extend this definition to any prime multiple of 'n'

For example 385 = 11 x 7 x 5 (expressed as the product of three prime numbers) therefore the numbers less than 385 relative prime to 385 ≡ 10 x 6 x 4 = 240

(12) Modulus: Modulus is the clock arithmetic, which reflects the remainder of two numbers or polynomials. Modulus can be positive or negative e.g. 15 mod 20 = 5 while 25 mod 20 = -5

I do not open up the truth to one who is not eager to get knowledge..................When I have presented one corner of subject to anyone, and he cannot from it learn other three, I do not repeat the message.

-Confucius (551-479 B.C.)

After word

We learn by doing and we learn from our errors. I hope that with this book you will attain self-mastery and self confidence of the subject and your errors will be minimal. With little discipline, anybody with the proficiency in mathematics with the help of this book, can attain leadership position in the digital age

By doing it by yourself shall unleash your creativity and it shall impart you whole new perspective. As you solve the problem, refer to this manual for clarifications.

Good luck,

Daljit S. Jandu

182-Riemann Hypothesis and PNT

Exercise is the best instrument in learning
- ROBERT RECORDE, The Whetstone of Witte (1557)

Two hours' daily exercise......will be enough
  to keep the hack fit for his work.
- M.H. MOHAN, The Handy Horse Book (1865)

Obey the scriptures until you are strong enough to do without them.

  - Swami Vivekanand
  tr. Christopher Isherwood, ed.
  Introduction of Vedas to Western World

The seer renounces fruits of his action and so reaches enlightenment.

  - Bhagavad Gita (6th Century B.C.)
  2nd tr. Swami Prashavananda and Christopher Isherwood, 1954

INDEX

Knowledge is of two kinds. We know the subject ourselves, or we know where we can find information on it.

- Samuel Johnson

A

Abraham Lincoln · 19
Alaine Connes · 123
Alan Turing · 123
Albert Bloch · 113
Albert Einstein · 19, 60, 66, 83,123,156
Alfred Noble Whitehead (1861-1947) · 85
algebraic computations · 114
algebraic equation · 158
algebraic extensions · 36
algorithm · 9, 25,149, 150
alternate positive and negative signs. · 48
alternate positive and negative terms ·51,52
alternate sign series · 39
American Mathematical Society; · 10
analytical continuation · 120, 121
analytical number theory. · 125
Andre Weil · 123
Andrew Odlyzko · 123
Apery · 8, 32, 174
Apery constant ·8,32, 122, 137, 140, 143, 174,175
Apery Constant · 50
arithmetic · 77, 127
asymptotic · 97, 109, 110, 170
Atle Selberg, · 123

B

B. Riemann · 138
Basic C++, · 148
Benjamin Franklin (1706-1790) · 172
Bernoulli numbers · 9,37,38,39,42, 116, 119,131, 136, 143, 152,153,157, 160,176
Bhagavad Gita (6th. Century B.C.) · 182
binary · 149
Binomial coefficient · 108,153

bio-informatics · 148
Block · 152
Bohr · 123

Borland Compiler · 148
boundary conditions · 30, 55, 56, 122, 130, 137, 140, 143, 156, 171, 175, 178
boundary equations · 82
boundary value · 55, 68, 80, 138

C

C++ · 147, 148
Carl Siegel · 123
Cartesian coordinate system · 157, 158, 161
cartesian plane · 31
Catalan's constant · 157,158
Central Processing Unit · 149
Charles de La Vallèe Poussin · 97,123
Chebyshev · 113, 125
Chinese Remainder Theorem · 23, 80, 91,146
circumference · 33
closing values · 127, 140
Code Warriors · 148
coefficient · 34
combinatorial · 55
compiler · 147, 148
complementary · 28, 47
complementary functions · 46
complex number ·9,131, 107, 112, 119, 124
complex variable · 126
complex wave equation · 61
computation · 31, 85, 133,147
Confucius (551-479 B.C.) · 89,180
conjecture · 42, 124,136, 138
conjugate complex roots · 143
Constant of Integration · 110
Convention · 60, 74
conventional mathematics · 93
convergence · 9,22,27,99,116,120
converging infinite series · 93
coordinate system · 109
correction factor · 88
correlate · 65
cosh(x) · 109

Cosine Integral · 95
CPAN · 147
critical line · 143
cubic series · 48
cyclicity · 23

D

data communication · 62
deep mathematical properties · 99
derivative · 110
Dirichlet · 113, 119, 127
discreet logarithmic function · 155
Discreet Logarithmic problem · 122
discreet wave type. · 61
divergence ·21,97, 116,127

E

$E = mc^2$ · 66
Edmund Landau, · 123
Edward Stillingfleet · 71
electrical induction · 9
Elizade · 152
elliptical function. · 157
emacs · 148
Emil Artin · 123
empirical facts · 41
equation · 26
Ernst Lindelof · 123
error correction ·70, 98, 126
error factor · 70
error term · 98
Euler · 77, 108, 113, 115, 124, 125
Euler extension numbers · 161
Euler Maclaurin · 141, 142
Euler numbers · 49,51, 157, 160
Euler totient function · 17
Euler's · 7, 43, 44, 62, 168
Euler's constant · 15, 135
Euler's extension · 9, 119
Euler's extension number · 40,143,176,177
Euler's famous equation · 140

Euler's famous result · 33,41,73, 114,130
Euler's Maclaurin · 143
Euler's number · 49,176
Euler's Totient Function · 146,179
even powered odd numbered series · 47
even powered sequential series · 53
Even zetas ·14,27, 174
exponent · 150
Exponential Integral · 94
extension · 9
extreme boundary · 29

F

Factorial · 34, 114, 179
Factorial (or Remainder) theorem · 46
factorial form · 24
factorial function · 119
factorial theorem · 176
Fermat's Little theorem. ·27,89, 146
floating point · 85, 150
floating point numbers · 147
Fourier series. · 7
Fourier Transformation · 30
Fran Libowitz · 107
Freeman Dyson · 124
full boundary conditions · 171
functional equation · 122, 125, 130, 136, 137, 138,139, 140, 171
fundamental theorem · 127
fundamental wave equation · 112

G

G.F. Gauss · 113
G.H. Hardy · 123
gamma function · 119, 120,121, 122, 129, 130, 134,135,136,137,168,170
Gauss · 77, 125, 140
Gauss conjecture · 67

Gauss Reciprocity theorem · 122
Gauss Theorem · 15, 77, 89,178
Gauss's Lemma · 172
Gauss's theorem · 140
Gaussian · 145
genus function · 33,39
geometric · 81
George Polya · 123
global consideration · 56
GNU foundation · 148
GNU scientific library · 148
Gram · 123,141
Greatest common divisor· 17
Guass · 124

H

Hadamard · 138, 141
Hadamard Product Formula: · 138
half boundary condition · 171,172
Harald Cramer · 123
harmonic series · 21, 116
hexadecimal, · 149
Hugh Montgomery. · 124
hyperbolic functions · 109,110
hyperbolic sine · 110
Hypothesis · 2,26,141

I

infinite · 9, 28
infinite degree · 43,140
infinite power · 37
infinite product · 7
infinite series · 175, 176, 177
infinitude combinations · 55
infinity **even** powers · 115
Information · 11
ingenuity · 61
Integers · 67
Integration formula · 100
internal memory · 147
irrational · 134

J

J. E. Littlewood · 123
Jacques Hadamard · 123
James Hopwood Jeans · 93
Java Big integer · 147
Java Big Integer · 147
Jorgen Pedersen · 123

K

Kronecker Delta · 152

L

large floating point computations · 147
Large Integer · 147
large processors · 88
lattice point · 178
law of nature · 130
laws of nature · 122, 172
least residue · 74, 172
Least Residue: · 74
Legendre · 77, 97,113, 120, 124, 128
Legendre Symbol · 74
Lejeune Dirichlet: · 35
Leopard Kroenecker · 67, 107
Lepowski · 152
$li(x)$ · 97
$Li(x)$ · 96, 97, 145, 164
Littlewood · 98
logarithm · 69
logarithmic differentiation · 132
logarithmic function · 99
Logarithmic Integral Function · 96

M

M.H. MOHAN, The Handy Horse Book (1865) · 182

Magnum Opus · 8
Mahatma Gandhi · 19
mapped to unity · 65
mathematical · 7
mathematics literature · 122
MHz. · 149
Microsoft Visual · 148
$Mli(x)$ · 146
Mobius function · 153
modulus · 80,73,75, 156,180
mutually prime · 91
mystery · 22, 52, 53,79
mystical · 83

N

N(T) · 138
Natural Numbers · 18
negative feedback · 178
Negative Infinite · 14
negative infinite product · 122, 140,173
negative least residue · 80
non- sinusoidal periodic function · 30
non-trivial roots · 31,127
normalizing · 25
number theory · 77, 107
numeric · 9, 61, 178
numeric algorithms · 62
Numeric Wave Equation. ·65,68, 112
numerical analysts · 77

O

octal · 149
odd powered all positive terms series · 49
odd powered and odd numbered series. · 52
odd powered infinite series · 49
odd powered odd numbered series, · 47
odd powered odd series · 49

odd zeta function. ·146, 156
odd zetas · 14,31,42,137, 140, 174, 175,176
orthogonal function · 159

P

periodicity · 50
PERL · 147
PERL Big Integer · 147
phase difference · 47
Pierre Deligne · 123
polynomial · 43, 45, 130, 140,180
polynomial function v · 15
positive and negative signs · 79
positive feedback · 178
positive infinite product · 122, 140, 156,173
positive integer · 49, 89, 121
premise · 24
prime · 80,89
prime counting · 126
Prime Counting Function · 99
Prime Number Theorem. · 65, 88, 99
prime numbers ·70, 80, 117, 122, 124, 140, 171,172, 180
probability · 35
purely cyclic · 146

Q

quadratic reciprocity theorem · 82, 122,140
quadratic residue · 73, 75, 79, 171
quantum system. · 30
Queen of Mathematics · 77

R

Ramanujan · 8, 22, 153, 155, 156
random numbers · 35
random positive · 31

re normalization · 139
real analysis · 127
Real Numbers · 18,128
real part · 31
real world applications · 93
reciprocal power, · 21
reciprocity · 79,156, 171, 172
relationship · 116
Remainder Theorem · 34, 90, 114, 158,179
renormalization · 152
residues · 65,135,177
Riemann · 2, 30, 98, 119, 122, 125, 137, 138, 139, 167
Riemann Hypothesis · 11, 27, 62, 65, 145, 172
Riemann Zeta function · 16,68, 122, 125, 139,143 ,151,171
Riemann's functional equation · 135
ROBERT RECORDE, · 182
roots · 114

S

S. Ramanujan · 123, 146
scholar · 89
Schrödinger Wave Equation · 61
sequential · 20
Series Expansion · 100
sh(x) · 109
sign bit · 150
simple poles · 121
Sine Integral · 94
sine wave equation · 62
Sir Isaac Newton · 7
Sir Winston Churchill · 167
Srinivas Ramanujan (1887-1920) · 96,153
Stirling Asymptotic Series · 170
straight positive sign · 50
subtly · 26
summation · 51
Sun-microsystem · 147
Swami Vivekanand · 33, 182

T

T.H. Huxley · 103
Taylor · 157
Technology. · 11
The positive as well as negative infinite product · 174
Thomas Hobbes · 77
Thomas Jefferson · 71
trigonometric functions · 43
trivial zeros · 26, 27, 130, 136, 137

U

unilateral · 122, 129
unilateral positive · 32
universal constant · 22
Unix · 148
Unix workstation · 148

V

Vallée Poussin · 141
virtual axis. · 109

Von Mangold Function · 164
Von Mangoldt · 123,138

W

Walter Isaacman (1952-   ) · 145
wave equation · 61, 65
William Wood worth · 60
Wilson Theorem · 23, 27, 89,146
windows operating systems. · 148

Z

zeta function · 116, 119, 124, 127, 143, 152, 175
zeta functional equation · 133

190-Riemann Hypothesis and PNT

192-Riemann Hypothesis and PNT